ARCO

MECHANICAL APTITUDE and SPATIAL RELATIONS TESTS

4th Edition

Joan U. Levy, Ph.D
Norman Levy, Ph.D

MACMILLAN • USA

Fourth Edition

Macmillan General Reference
A Simon & Schuster Macmillan Company
1633 Broadway
New York, NY 10019-6785

Macmillan Publishing books may be purchased for business or sales promotional use. For information please write: Special Markets Department, Macmillan Publishing USA, 1633 Broadway, New York, NY 10019-6785.

An ARCO Book

ARCO is a registered trademark of Simon & Schuster, Inc.
MACMILLAN is a registered trademark of Macmillan, Inc.

Library of Congress Number: 98-88657

ISBN: 0-02-862816-0

Manufactured in the United States of America

10 9 8 7 6 5 4 3 2 1

TABLE OF CONTENTS

PART THREE. MECHANICAL APTITUDE AND ACHIEVEMENT

For Joshua Seth
and
Jessica Dawn

two of the best reasons for our writing
this book.

ACKNOWLEDGMENT

The study of tools, their usage and the practical experience of creating with them is becoming a dying art. The authors gratefully acknowledge the efforts of Howard Levy, of Marietta, Georgia, for his research and contribution to this publication.

WHAT THIS BOOK WILL DO FOR YOU

Testing is a big business. Whether you are applying for a training program or seeking a job in the civil service, private industry or the military, chances are that you will have to take a test to demonstrate your aptitude for the work involved. If the job is in the field of mechanical work, the test you take will probably be one of the many tests designed to measure mechanical aptitude and spatial relations. This book presents hundreds of sample questions from such tests. By showing you the kind of test you will have to take, it will give you the confidence you need to do your best.

Arco Publishing Inc. has followed testing trends and methods ever since the firm was founded in 1937. We specialize in books that prepare people for tests. Based on this experience, we have prepared the best possible book to help you score high.

This book will tell you exactly what to study by presenting in full every type of question asked on Mechanical Aptitude and Spatial Relations tests.

This book will help you find your weaknesses. Once you know what subjects you're weak in, you can concentrate your study on those areas that need the most work. This kind of selective study yields maximum test results.

This book will give you the *feel* of the exam. Almost all our practice questions are taken from actual previous exams. On the day of your test, you'll see how closely this book follows the format of the real test.

This book will give you confidence *now*, while you are preparing for the test. It will build your self-confidence as you proceed and will prevent the kind of test anxiety that causes low test scores.

WHAT IS MECHANICAL APTITUDE?

Industry, government, schools, and the military all use tests to find those applicants who are most likely to succeed and then to place them where they will work best. "Does this individual have the potential for doing well with us?" is what the employer wants to know. "Does he or she, in addition to special skills or training, have an *aptitude* for the job?"

Psychologists have discovered that success in a great variety of occupations and fields of study can be predicted with mechanical aptitude and spatial relations tests. As a result, mechanical aptitude and achievement tests are used for selection and classification purposes in schools, industry, government, and the armed services. Sometimes they measure your ability to use your hands or to manipulate materials. Other times they measure your dexterity or your ability to visualize shapes and forms and to move them around in different patterns. Some people are better at the tasks than others, even without any training. It is believed that such people have better mechanical "aptitude."

Frequently tests that are designed to measure mechanical aptitude draw upon a person's acquired knowledge in mechanical matters. The theory here is that those who are likely to do well in mechanical work have interested themselves in it and have absorbed more from what they have experienced than those who are likely to fail in such work.

Mechanical aptitude tests are given in many forms. The questions may be entirely verbal. Or they may involve the use of diagrams and pictures. They may be given orally, or they may be printed and require written answers. And sometimes the candidate may be asked to work with his hands on special testing equipment while the examiner observes and times his effort. Another form of manual test consists of actual performance on a job, or in the production of *standardized worksamples*. In such tests the work set for the candidate if very much like the work required by the job. For economy and speed, worksamples are frequently presented on simulated or miniaturized equipment.

1

MECHANICAL APTITUDE AND SPATIAL RELATIONS: WHAT'S THE DIFFERENCE?

It would be well at this point to discuss the differences and the similarities between mechanical aptitude and spatial relations tests. Many factors go into making a person successful in his work. One of the most important of these in many (but not all) cases is mechanical aptitude. But mechanical aptitude is itself a complex function, the sum of several different capacities, one of which is the ability to perceive spatial relations. It has been found that spatial ability plays an extremely important part in performing certain jobs—patternmaking, drafting, engineering, and dentistry, for example—while in other jobs such as inspecting, assembling, and tool and die making, other mechanical abilities are dominant. Because research has shown that spatial ability is, for *some* jobs, the most vital part of mechanical aptitude, spatial relations tests are sometimes given separately. In other cases, spatial relations occurs as part of a larger mechanical aptitude test. And in still other cases, a mechanical aptitude test may have no spatial relations section at all. It all depends on the purpose of the test and the nature of the job. By definition spatial relations tests are *not* knowledge tests, whereas mechanical aptitude tests frequently demand some knowledge of a trade or a familiarity with what might be called "mechanical information of everyday life."

HOW TO USE THIS BOOK

This book is divided into three parts, corresponding to the three areas most likely to be covered on tests of mechanical aptitude—Spatial Relations, Reasoning with Symbols, and Mechanical Aptitude and Achievement. Each part provides a complete explanation of and comprehensive practice with every type of question currently used in civil service, military, and industry tests for selection and placement of entry-level applicants and trainees in mechanical work.

SPATIAL RELATIONS TESTS

Studies have shown spatial relations tests to be especially valuable in predicting success in protective and service occupations, mechanical repair, electrical, structural, and processing work, operation of complex machines, and gross manual work. Although they are called "aptitude" tests, meaning tests of natural ability or capacity to learn, spatial relations tests have been show to respond positively to increased familiarity with the tests and practice with the kinds of questions asked.

Part One of this book presents a variety of Spatial Relations Tests designed to measure all of the following functions:

Spatial Orientation, which means the ability to comprehend static or stationary objects and is measured by Hidden Figures, Matching Parts and Figures, Spatial Views, Cube Counting, and Touching Cube Tests.

Spatial Visualization, or the ability to picture the way an object will look after it has been moved to another position, which is measured by Pattern Analysis, Solid Figure Turning and Cube Turning Tests.

Perceptual Ability, which is the ability to compare, contrast and rank similar figures and is measured by Similarities and Differences Between Objects, Line Ranking, and Angle Comparison Tests.

Visual-Motor Coordination, or the ability to perform tasks with speed and accuracy, which is measured by Letter-Symbol Coding, Inspection, Examining Objects, and Pursuit Tests.

Often the most difficult part of Spatial Relations Tests is understanding and following the directions for each test. Practice with the wide variety of tests in Part One will make it easier for you to perform well no matter what kind of Spatial Relations Test you may have to take.

SYMBOL REASONING TESTS

Another aspect of Spatial and Mechanical Aptitude is the ability to reason with symbols. Symbol Reasoning Tests provide a means of measuring your capacity for understanding and following directions and for thinking logically. By using symbols rather than words, these tests can measure learning ability independent of language and reading skills, thereby providing a fair estimate of the ability of individuals from diverse cultural and educational backgrounds.

Symbol Reasoning Tests require you to spot underlying relationships or common characteristics and to pinpoint patterns of progression of abstract symbols, letters, or numbers. These abilities have been found useful in screening applicants for entry-level jobs as factory workers, heavy equipment operators, service technicians, keypunch operators, and computer programmers.

Your ability to reason with symbols may be measured by Symbol Series, Symbol Analogies, Figure Classification, and Series Reasoning Tests such as the one found in Part Two of this book.

MECHANICAL APTITUDE TESTS

Mechanical Aptitude Tests are useful in predicting success in hundreds of civil service and private industry jobs requiring the ability to operate, service, or maintain machinery. Over the years, personnel experts and psychologists have worked together to come up with a series of paper and pencil tests that will accurately predict the ability of a person to han-

dle mechanical work. The tests in Part Three of this book provide a sampling of the results of these efforts.

In practice, it has been found that persons who do well on Tool Recognition, Mechanical Insight, Mechanical Knowledge, and Shop Arithmetic Tests are likely to do well in the workshop too. Practice with the sample tests in these areas will enable you to discover your weak points before you take your examination and to use your study time more effectively. If you have difficulty with the mechanical concepts, you may do well to review basic physics. If the calculations stump you, brush up on arithmetic computation. Your efforts will be rewarded with higher test scores and a better chance of getting the job you want.

In using this book it would be wise to keep in mind that there is no one ability that accounts for success or can be used to predict success in a given job or field of study. However, psychologists have developed formulas by which they can take *all* the individual aptitudes into account. Because such methods have proved valuable in employee and student selection, it will be to your advantage to work diligently on every section of this book, devoting the greatest time and effort to those areas in which you find yourself to be weak. The effect of all this practice will be evident on any Mechanical or Spatial Aptitude Test you may have to take.

TECHNIQUES OF STUDY AND TEST-TAKING

Although a thorough knowledge of the subject matter is the most important factor in succeeding on your exam, the following suggestions could raise your score substantially. These few pointers will give you the strategy employed on tests by those who are most successful in this not-so-mysterious art. It's really quite simple. Do things right . . . right from the beginning. Make these successful methods a habit. Then you'll get the greatest dividends from the time you invest in this book.

HOW TO PREPARE FOR YOUR EXAM

1. Budget your time. Set aside definite hours each day for concentrated study. Keep to your schedule.

2. Study alone. You will concentrate better when you work by yourself. Keep a list of questions you cannot answer and points you are unsure of to talk over with a friend who is preparing for the same exam. Plan to exchange ideas at a joint review session just before the test.

3. Eliminate distractions. Disturbances caused by family and neighbor activities (telephone calls, chit-chat, TV programs, etc.) work to your disadvantage. Try to find a quiet, private place for your study.

4. Use the library. Most communities have excellent library facilities. The library is free from those distractions that make it difficult to study at home. Moreover, if you need to do any additional research in your subject area, the library can provide more study material than you have at home.

5. Answer all the questions in this book. Don't be satisfied just to know the correct answer to each question. Try to understand what makes each incorrect answer wrong. If you understand the reasoning behind each practice question, you will be better prepared for the questions on the "real" exam.

6. Get the "feel" of the exam. The sample questions in this book will give you that "feel." These questions cover just about every kind of mechanical aptitude and spatial relations test currently in use.

7. Take the Practice Tests as "real" tests. Work through each test without any interruptions. Do not spend too much time on any one question. If a question seems too difficult, go to the next one. If time permits, go back to the omitted question.

8. Tailor your study to the subject matter. Skim or scan. Don't study everything in the same manner. Obviously, certain areas are more important than others.

9. Organize yourself. When you sit down to study, make sure you have everything you need. That way you won't have to waste valuable study time looking around for a pencil or scratch paper.

10. Keep physically fit. You cannot retain information well when you are uncomfortable, headachy, or tense. Physical health promotes mental efficiency.

HOW TO TAKE AN EXAM

1. Get to the examination room about ten minutes ahead of time. You'll get a better start when you are accustomed to the room. If the room is too cold or too warm, or not well ventilated, call these conditions to the attention of the person in charge.

2. Make sure that you read the instructions carefully. In most cases, test-takers lose credit because they misread some important point in the directions. For example, they may select the incorrect choice when the directions specify selecting the correct choice.

3. Be confident. Statistics conclusively show that you are more likely to get high scores if you have prepared for your test. It is important to know that you are not expected to answer every question correctly. The questions usually have a range of difficulty and differentiate between several levels of skill.

4. Skip hard questions and go back later. It is a good idea to make a mark in your test booklet next to each question you cannot answer easily; then you can go back to those questions later, after you've answered all the questions you know. Don't panic if you cannot answer a question. Keep calm and move on. Usually the easier questions are presented at the beginning of the exam and the questions become gradually more difficult.

If you do skip ahead on the exam, be sure to skip ahead also on your answer sheet. A good technique is periodically to check the number of the question on the answer sheet with the number of the question on the test. You should do this every time you decide to skip a question. If you fail to skip the corresponding answer blank for that question, all of your following answers will be wrong.

Each student is stronger in some areas than in others. No one is expected to know all the answers. Do not waste time agonizing over a difficult question because it may keep you from getting to other questions that you can answer correctly.

5. Guess if you are not sure. No penalty is given for guessing when these exams are scored. Therefore, it is better to guess than to omit an answer.

6. Mark the answer sheet clearly. When you take the examination, you will mark your answers to the multiple-choice questions on a separate answer sheet that will be given to you at the test center. If you have not worked with an answer sheet before, it is in your best interest to become familiar with the procedures involved. Remember, knowing the correct answer is not enough! If you do not mark the sheet correctly, so that it can be machine-scored, you will not get credit for your answers!

In addition to marking answers on the separate answer sheet, you will be asked to give your name and other information, including your social security number. As a precaution, bring along your social security number for identification purposes.

Read the directions carefully and follow them exactly. If they ask you to print your name in the boxes provided, write only one letter in each box. If your name is longer than the number provided, omit the letters that do not fit. Remember, you are writing for a machine; it does not have judgment. It can only record the pencil marks you make on the answer sheet.

Use the answer sheet to record all your answers to questions. Each question, or item, has four or five answer choices labeled (A), (B), (C), (D), (E). You will be asked to choose the letter that stands for the best answer. Then you will be asked to mark your answer by blackening the appropriate space on your answer sheet. Be sure that each space you choose and blacken with your pencil is completely blackened. The machine will "read" your answers in terms of spaces blackened. Make sure that only one answer is clearly blackened. If you erase an answer, erase it completely and mark your new answer clearly. The machine will give you credit only for clearly marked answers. It does not pause to decide whether you really meant (B) or (C).

Make sure that the number of the question you are being asked on the question sheet corresponds to the number of the question you are answering on the answer sheet. It is a good idea to check the numbers of questions and answers frequently. If you decide to skip a question, but fail to skip the corresponding answer blank for that question, all your answers after that will be wrong.

7. Read each question carefully. The exam questions are not designed to trick you through misleading or ambiguous alternative choices. On the other hand, they are not all direct questions of factual information. Some are designed to elicit responses that reveal your ability to reason, or to interpret a fact or idea. It's up to you to read each question carefully so you know what is being asked. The exam authors have tried to make the questions clear. Do not go astray looking for hidden meanings.

8. Don't answer too fast. The multiple-choice questions that you will meet are not superficial exercises. They are designed to test not only your memory but also your understanding and insight. Do not place too much emphasis on speed. The time element is a factor, but it is not all-important. Accuracy should not be sacrificed for speed.

9. Know the rules of the test center. You need to bring with you to the test center your Admission Form, your social security number, and several No. 2 pencils. Arrive on time, as you may not be admitted after testing has begun. Instructions for taking the test will be read to you by the test supervisor and time will be called when the test is over. If you have questions, you may ask them of the supervisor. Do not give or receive assistance while taking the exams. If you do, you will be asked to turn in all test materials and told to leave the room. You will not be permitted to return and your tests will not be scored.

Part One
Spatial Relations

PART ONE

SPATIAL RELATIONS
Answer Sheets

To simulate actual examination conditions, mark your answers to each question in this chapter on these answer sheets. Make one clear, black mark for each answer. If you decide to change an answer, erase your error completely. On machine-scored examinations, additional or stray marks on your answer sheet may be counted as mistakes. After you have taken all the tests, compare your answers to the Correct Answers at the end of the chapter to see where you stand.

HIDDEN FIGURES

Hidden Figures—Test I

1 Ⓐ Ⓑ Ⓒ Ⓓ Ⓔ	8 Ⓐ Ⓑ Ⓒ Ⓓ Ⓔ	15 Ⓐ Ⓑ Ⓒ Ⓓ Ⓔ	22 Ⓐ Ⓑ Ⓒ Ⓓ Ⓔ
2 Ⓐ Ⓑ Ⓒ Ⓓ Ⓔ	9 Ⓐ Ⓑ Ⓒ Ⓓ Ⓔ	16 Ⓐ Ⓑ Ⓒ Ⓓ Ⓔ	23 Ⓐ Ⓑ Ⓒ Ⓓ Ⓔ
3 Ⓐ Ⓑ Ⓒ Ⓓ Ⓔ	10 Ⓐ Ⓑ Ⓒ Ⓓ Ⓔ	17 Ⓐ Ⓑ Ⓒ Ⓓ Ⓔ	24 Ⓐ Ⓑ Ⓒ Ⓓ Ⓔ
4 Ⓐ Ⓑ Ⓒ Ⓓ Ⓔ	11 Ⓐ Ⓑ Ⓒ Ⓓ Ⓔ	18 Ⓐ Ⓑ Ⓒ Ⓓ Ⓔ	25 Ⓐ Ⓑ Ⓒ Ⓓ Ⓔ
5 Ⓐ Ⓑ Ⓒ Ⓓ Ⓔ	12 Ⓐ Ⓑ Ⓒ Ⓓ Ⓔ	19 Ⓐ Ⓑ Ⓒ Ⓓ Ⓔ	26 Ⓐ Ⓑ Ⓒ Ⓓ Ⓔ
6 Ⓐ Ⓑ Ⓒ Ⓓ Ⓔ	13 Ⓐ Ⓑ Ⓒ Ⓓ Ⓔ	20 Ⓐ Ⓑ Ⓒ Ⓓ Ⓔ	27 Ⓐ Ⓑ Ⓒ Ⓓ Ⓔ
7 Ⓐ Ⓑ Ⓒ Ⓓ Ⓔ	14 Ⓐ Ⓑ Ⓒ Ⓓ Ⓔ	21 Ⓐ Ⓑ Ⓒ Ⓓ Ⓔ	28 Ⓐ Ⓑ Ⓒ Ⓓ Ⓔ

Hidden Figures—Test II

1 Ⓐ Ⓑ Ⓒ Ⓓ Ⓔ	9 Ⓐ Ⓑ Ⓒ Ⓓ Ⓔ	17 Ⓐ Ⓑ Ⓒ Ⓓ Ⓔ	25 Ⓐ Ⓑ Ⓒ Ⓓ Ⓔ
2 Ⓐ Ⓑ Ⓒ Ⓓ Ⓔ	10 Ⓐ Ⓑ Ⓒ Ⓓ Ⓔ	18 Ⓐ Ⓑ Ⓒ Ⓓ Ⓔ	26 Ⓐ Ⓑ Ⓒ Ⓓ Ⓔ
3 Ⓐ Ⓑ Ⓒ Ⓓ Ⓔ	11 Ⓐ Ⓑ Ⓒ Ⓓ Ⓔ	19 Ⓐ Ⓑ Ⓒ Ⓓ Ⓔ	27 Ⓐ Ⓑ Ⓒ Ⓓ Ⓔ
4 Ⓐ Ⓑ Ⓒ Ⓓ Ⓔ	12 Ⓐ Ⓑ Ⓒ Ⓓ Ⓔ	20 Ⓐ Ⓑ Ⓒ Ⓓ Ⓔ	28 Ⓐ Ⓑ Ⓒ Ⓓ Ⓔ
5 Ⓐ Ⓑ Ⓒ Ⓓ Ⓔ	13 Ⓐ Ⓑ Ⓒ Ⓓ Ⓔ	21 Ⓐ Ⓑ Ⓒ Ⓓ Ⓔ	29 Ⓐ Ⓑ Ⓒ Ⓓ Ⓔ
6 Ⓐ Ⓑ Ⓒ Ⓓ Ⓔ	14 Ⓐ Ⓑ Ⓒ Ⓓ Ⓔ	22 Ⓐ Ⓑ Ⓒ Ⓓ Ⓔ	30 Ⓐ Ⓑ Ⓒ Ⓓ Ⓔ
7 Ⓐ Ⓑ Ⓒ Ⓓ Ⓔ	15 Ⓐ Ⓑ Ⓒ Ⓓ Ⓔ	23 Ⓐ Ⓑ Ⓒ Ⓓ Ⓔ	31 Ⓐ Ⓑ Ⓒ Ⓓ Ⓔ
8 Ⓐ Ⓑ Ⓒ Ⓓ Ⓔ	16 Ⓐ Ⓑ Ⓒ Ⓓ Ⓔ	24 Ⓐ Ⓑ Ⓒ Ⓓ Ⓔ	32 Ⓐ Ⓑ Ⓒ Ⓓ Ⓔ

Hidden Figures—Test III

1 Ⓐ Ⓑ Ⓒ Ⓓ Ⓔ	9 Ⓐ Ⓑ Ⓒ Ⓓ Ⓔ	17 Ⓐ Ⓑ Ⓒ Ⓓ Ⓔ	25 Ⓐ Ⓑ Ⓒ Ⓓ Ⓔ
2 Ⓐ Ⓑ Ⓒ Ⓓ Ⓔ	10 Ⓐ Ⓑ Ⓒ Ⓓ Ⓔ	18 Ⓐ Ⓑ Ⓒ Ⓓ Ⓔ	26 Ⓐ Ⓑ Ⓒ Ⓓ Ⓔ
3 Ⓐ Ⓑ Ⓒ Ⓓ Ⓔ	11 Ⓐ Ⓑ Ⓒ Ⓓ Ⓔ	19 Ⓐ Ⓑ Ⓒ Ⓓ Ⓔ	27 Ⓐ Ⓑ Ⓒ Ⓓ Ⓔ
4 Ⓐ Ⓑ Ⓒ Ⓓ Ⓔ	12 Ⓐ Ⓑ Ⓒ Ⓓ Ⓔ	20 Ⓐ Ⓑ Ⓒ Ⓓ Ⓔ	28 Ⓐ Ⓑ Ⓒ Ⓓ Ⓔ
5 Ⓐ Ⓑ Ⓒ Ⓓ Ⓔ	13 Ⓐ Ⓑ Ⓒ Ⓓ Ⓔ	21 Ⓐ Ⓑ Ⓒ Ⓓ Ⓔ	29 Ⓐ Ⓑ Ⓒ Ⓓ Ⓔ
6 Ⓐ Ⓑ Ⓒ Ⓓ Ⓔ	14 Ⓐ Ⓑ Ⓒ Ⓓ Ⓔ	22 Ⓐ Ⓑ Ⓒ Ⓓ Ⓔ	30 Ⓐ Ⓑ Ⓒ Ⓓ Ⓔ
7 Ⓐ Ⓑ Ⓒ Ⓓ Ⓔ	15 Ⓐ Ⓑ Ⓒ Ⓓ Ⓔ	23 Ⓐ Ⓑ Ⓒ Ⓓ Ⓔ	31 Ⓐ Ⓑ Ⓒ Ⓓ Ⓔ
8 Ⓐ Ⓑ Ⓒ Ⓓ Ⓔ	16 Ⓐ Ⓑ Ⓒ Ⓓ Ⓔ	24 Ⓐ Ⓑ Ⓒ Ⓓ Ⓔ	32 Ⓐ Ⓑ Ⓒ Ⓓ Ⓔ

MATCHING PARTS AND FIGURES

Matching Parts and Figures—Test I

1 Ⓐ Ⓑ Ⓒ Ⓓ Ⓔ	4 Ⓐ Ⓑ Ⓒ Ⓓ Ⓔ	7 Ⓐ Ⓑ Ⓒ Ⓓ Ⓔ	10 Ⓐ Ⓑ Ⓒ Ⓓ Ⓔ	13 Ⓐ Ⓑ Ⓒ Ⓓ Ⓔ
2 Ⓐ Ⓑ Ⓒ Ⓓ Ⓔ	5 Ⓐ Ⓑ Ⓒ Ⓓ Ⓔ	8 Ⓐ Ⓑ Ⓒ Ⓓ Ⓔ	11 Ⓐ Ⓑ Ⓒ Ⓓ Ⓔ	14 Ⓐ Ⓑ Ⓒ Ⓓ Ⓔ
3 Ⓐ Ⓑ Ⓒ Ⓓ Ⓔ	6 Ⓐ Ⓑ Ⓒ Ⓓ Ⓔ	9 Ⓐ Ⓑ Ⓒ Ⓓ Ⓔ	12 Ⓐ Ⓑ Ⓒ Ⓓ Ⓔ	15 Ⓐ Ⓑ Ⓒ Ⓓ Ⓔ

Matching Parts and Figures—Test II

1 Ⓐ Ⓑ Ⓒ Ⓓ Ⓔ	4 Ⓐ Ⓑ Ⓒ Ⓓ Ⓔ	7 Ⓐ Ⓑ Ⓒ Ⓓ Ⓔ	10 Ⓐ Ⓑ Ⓒ Ⓓ Ⓔ	13 Ⓐ Ⓑ Ⓒ Ⓓ Ⓔ
2 Ⓐ Ⓑ Ⓒ Ⓓ Ⓔ	5 Ⓐ Ⓑ Ⓒ Ⓓ Ⓔ	8 Ⓐ Ⓑ Ⓒ Ⓓ Ⓔ	11 Ⓐ Ⓑ Ⓒ Ⓓ Ⓔ	14 Ⓐ Ⓑ Ⓒ Ⓓ Ⓔ
3 Ⓐ Ⓑ Ⓒ Ⓓ Ⓔ	6 Ⓐ Ⓑ Ⓒ Ⓓ Ⓔ	9 Ⓐ Ⓑ Ⓒ Ⓓ Ⓔ	12 Ⓐ Ⓑ Ⓒ Ⓓ Ⓔ	15 Ⓐ Ⓑ Ⓒ Ⓓ Ⓔ

Matching Parts and Figures—Test III

1 Ⓐ Ⓑ Ⓒ Ⓓ	3 Ⓐ Ⓑ Ⓒ Ⓓ	5 Ⓐ Ⓑ Ⓒ Ⓓ	7 Ⓐ Ⓑ Ⓒ Ⓓ
2 Ⓐ Ⓑ Ⓒ Ⓓ	4 Ⓐ Ⓑ Ⓒ Ⓓ	6 Ⓐ Ⓑ Ⓒ Ⓓ	8 Ⓐ Ⓑ Ⓒ Ⓓ

Matching Parts and Figures—Test IV

1 Ⓐ Ⓑ Ⓒ Ⓓ   4 Ⓐ Ⓑ Ⓒ Ⓓ   7 Ⓐ Ⓑ Ⓒ Ⓓ

2 Ⓐ Ⓑ Ⓒ Ⓓ   5 Ⓐ Ⓑ Ⓒ Ⓓ   8 Ⓐ Ⓑ Ⓒ Ⓓ

3 Ⓐ Ⓑ Ⓒ Ⓓ   6 Ⓐ Ⓑ Ⓒ Ⓓ   9 Ⓐ Ⓑ Ⓒ Ⓓ

SPATIAL VIEWS

Spatial Views—Test I

1 Ⓐ Ⓑ Ⓒ Ⓓ   3 Ⓐ Ⓑ Ⓒ Ⓓ   5 Ⓐ Ⓑ Ⓒ Ⓓ   7 Ⓐ Ⓑ Ⓒ Ⓓ   9 Ⓐ Ⓑ Ⓒ Ⓓ

2 Ⓐ Ⓑ Ⓒ Ⓓ   4 Ⓐ Ⓑ Ⓒ Ⓓ   6 Ⓐ Ⓑ Ⓒ Ⓓ   8 Ⓐ Ⓑ Ⓒ Ⓓ   10 Ⓐ Ⓑ Ⓒ Ⓓ

Spatial Views—Test II

1 Ⓐ Ⓑ Ⓒ Ⓓ   4 Ⓐ Ⓑ Ⓒ Ⓓ   7 Ⓐ Ⓑ Ⓒ Ⓓ

2 Ⓐ Ⓑ Ⓒ Ⓓ   5 Ⓐ Ⓑ Ⓒ Ⓓ   8 Ⓐ Ⓑ Ⓒ Ⓓ

3 Ⓐ Ⓑ Ⓒ Ⓓ   6 Ⓐ Ⓑ Ⓒ Ⓓ   9 Ⓐ Ⓑ Ⓒ Ⓓ

Spatial Views—Test III

1 Ⓐ Ⓑ Ⓒ Ⓓ   3 Ⓐ Ⓑ Ⓒ Ⓓ   5 Ⓐ Ⓑ Ⓒ Ⓓ   7 Ⓐ Ⓑ Ⓒ Ⓓ   9 Ⓐ Ⓑ Ⓒ Ⓓ

2 Ⓐ Ⓑ Ⓒ Ⓓ   4 Ⓐ Ⓑ Ⓒ Ⓓ   6 Ⓐ Ⓑ Ⓒ Ⓓ   8 Ⓐ Ⓑ Ⓒ Ⓓ   10 Ⓐ Ⓑ Ⓒ Ⓓ

CUBE COUNTING

Cube Counting—Test I

1. _____	7. _____	13. _____	19. _____	25. _____
2. _____	8. _____	14. _____	20. _____	26. _____
3. _____	9. _____	15. _____	21. _____	27. _____
4. _____	10. _____	16. _____	22. _____	28. _____
5. _____	11. _____	17. _____	23. _____	29. _____
6. _____	12. _____	18. _____	24. _____	30. _____

Cube Counting—Test II

1 Ⓐ Ⓑ Ⓒ Ⓓ Ⓔ	23 Ⓐ Ⓑ Ⓒ Ⓓ Ⓔ	45 Ⓐ Ⓑ Ⓒ Ⓓ Ⓔ	67 Ⓐ Ⓑ Ⓒ Ⓓ Ⓔ
2 Ⓐ Ⓑ Ⓒ Ⓓ Ⓔ	24 Ⓐ Ⓑ Ⓒ Ⓓ Ⓔ	46 Ⓐ Ⓑ Ⓒ Ⓓ Ⓔ	68 Ⓐ Ⓑ Ⓒ Ⓓ Ⓔ
3 Ⓐ Ⓑ Ⓒ Ⓓ Ⓔ	25 Ⓐ Ⓑ Ⓒ Ⓓ Ⓔ	47 Ⓐ Ⓑ Ⓒ Ⓓ Ⓔ	69 Ⓐ Ⓑ Ⓒ Ⓓ Ⓔ
4 Ⓐ Ⓑ Ⓒ Ⓓ Ⓔ	26 Ⓐ Ⓑ Ⓒ Ⓓ Ⓔ	48 Ⓐ Ⓑ Ⓒ Ⓓ Ⓔ	70 Ⓐ Ⓑ Ⓒ Ⓓ Ⓔ
5 Ⓐ Ⓑ Ⓒ Ⓓ Ⓔ	27 Ⓐ Ⓑ Ⓒ Ⓓ Ⓔ	49 Ⓐ Ⓑ Ⓒ Ⓓ Ⓔ	71 Ⓐ Ⓑ Ⓒ Ⓓ Ⓔ
6 Ⓐ Ⓑ Ⓒ Ⓓ Ⓔ	28 Ⓐ Ⓑ Ⓒ Ⓓ Ⓔ	50 Ⓐ Ⓑ Ⓒ Ⓓ Ⓔ	72 Ⓐ Ⓑ Ⓒ Ⓓ Ⓔ
7 Ⓐ Ⓑ Ⓒ Ⓓ Ⓔ	29 Ⓐ Ⓑ Ⓒ Ⓓ Ⓔ	51 Ⓐ Ⓑ Ⓒ Ⓓ Ⓔ	73 Ⓐ Ⓑ Ⓒ Ⓓ Ⓔ
8 Ⓐ Ⓑ Ⓒ Ⓓ Ⓔ	30 Ⓐ Ⓑ Ⓒ Ⓓ Ⓔ	52 Ⓐ Ⓑ Ⓒ Ⓓ Ⓔ	74 Ⓐ Ⓑ Ⓒ Ⓓ Ⓔ
9 Ⓐ Ⓑ Ⓒ Ⓓ Ⓔ	31 Ⓐ Ⓑ Ⓒ Ⓓ Ⓔ	53 Ⓐ Ⓑ Ⓒ Ⓓ Ⓔ	75 Ⓐ Ⓑ Ⓒ Ⓓ Ⓔ
10 Ⓐ Ⓑ Ⓒ Ⓓ Ⓔ	32 Ⓐ Ⓑ Ⓒ Ⓓ Ⓔ	54 Ⓐ Ⓑ Ⓒ Ⓓ Ⓔ	76 Ⓐ Ⓑ Ⓒ Ⓓ Ⓔ
11 Ⓐ Ⓑ Ⓒ Ⓓ Ⓔ	33 Ⓐ Ⓑ Ⓒ Ⓓ Ⓔ	55 Ⓐ Ⓑ Ⓒ Ⓓ Ⓔ	77 Ⓐ Ⓑ Ⓒ Ⓓ Ⓔ
12 Ⓐ Ⓑ Ⓒ Ⓓ Ⓔ	34 Ⓐ Ⓑ Ⓒ Ⓓ Ⓔ	56 Ⓐ Ⓑ Ⓒ Ⓓ Ⓔ	78 Ⓐ Ⓑ Ⓒ Ⓓ Ⓔ
13 Ⓐ Ⓑ Ⓒ Ⓓ Ⓔ	35 Ⓐ Ⓑ Ⓒ Ⓓ Ⓔ	57 Ⓐ Ⓑ Ⓒ Ⓓ Ⓔ	79 Ⓐ Ⓑ Ⓒ Ⓓ Ⓔ
14 Ⓐ Ⓑ Ⓒ Ⓓ Ⓔ	36 Ⓐ Ⓑ Ⓒ Ⓓ Ⓔ	58 Ⓐ Ⓑ Ⓒ Ⓓ Ⓔ	80 Ⓐ Ⓑ Ⓒ Ⓓ Ⓔ
15 Ⓐ Ⓑ Ⓒ Ⓓ Ⓔ	37 Ⓐ Ⓑ Ⓒ Ⓓ Ⓔ	59 Ⓐ Ⓑ Ⓒ Ⓓ Ⓔ	81 Ⓐ Ⓑ Ⓒ Ⓓ Ⓔ
16 Ⓐ Ⓑ Ⓒ Ⓓ Ⓔ	38 Ⓐ Ⓑ Ⓒ Ⓓ Ⓔ	60 Ⓐ Ⓑ Ⓒ Ⓓ Ⓔ	82 Ⓐ Ⓑ Ⓒ Ⓓ Ⓔ
17 Ⓐ Ⓑ Ⓒ Ⓓ Ⓔ	39 Ⓐ Ⓑ Ⓒ Ⓓ Ⓔ	61 Ⓐ Ⓑ Ⓒ Ⓓ Ⓔ	83 Ⓐ Ⓑ Ⓒ Ⓓ Ⓔ
18 Ⓐ Ⓑ Ⓒ Ⓓ Ⓔ	40 Ⓐ Ⓑ Ⓒ Ⓓ Ⓔ	62 Ⓐ Ⓑ Ⓒ Ⓓ Ⓔ	84 Ⓐ Ⓑ Ⓒ Ⓓ Ⓔ
19 Ⓐ Ⓑ Ⓒ Ⓓ Ⓔ	41 Ⓐ Ⓑ Ⓒ Ⓓ Ⓔ	63 Ⓐ Ⓑ Ⓒ Ⓓ Ⓔ	85 Ⓐ Ⓑ Ⓒ Ⓓ Ⓔ
20 Ⓐ Ⓑ Ⓒ Ⓓ Ⓔ	42 Ⓐ Ⓑ Ⓒ Ⓓ Ⓔ	64 Ⓐ Ⓑ Ⓒ Ⓓ Ⓔ	86 Ⓐ Ⓑ Ⓒ Ⓓ Ⓔ
21 Ⓐ Ⓑ Ⓒ Ⓓ Ⓔ	43 Ⓐ Ⓑ Ⓒ Ⓓ Ⓔ	65 Ⓐ Ⓑ Ⓒ Ⓓ Ⓔ	87 Ⓐ Ⓑ Ⓒ Ⓓ Ⓔ
22 Ⓐ Ⓑ Ⓒ Ⓓ Ⓔ	44 Ⓐ Ⓑ Ⓒ Ⓓ Ⓔ	66 Ⓐ Ⓑ Ⓒ Ⓓ Ⓔ	88 Ⓐ Ⓑ Ⓒ Ⓓ Ⓔ

PATTERN ANALYSIS

Cardboard Folding—Test I

1 Ⓐ Ⓑ Ⓒ Ⓓ   4 Ⓐ Ⓑ Ⓒ Ⓓ   7 Ⓐ Ⓑ Ⓒ Ⓓ

2 Ⓐ Ⓑ Ⓒ Ⓓ   5 Ⓐ Ⓑ Ⓒ Ⓓ   8 Ⓐ Ⓑ Ⓒ Ⓓ

3 Ⓐ Ⓑ Ⓒ Ⓓ   6 Ⓐ Ⓑ Ⓒ Ⓓ   9 Ⓐ Ⓑ Ⓒ Ⓓ

Cardboard Folding—Test II

1 Ⓐ Ⓑ Ⓒ Ⓓ   4 Ⓐ Ⓑ Ⓒ Ⓓ   7 Ⓐ Ⓑ Ⓒ Ⓓ

2 Ⓐ Ⓑ Ⓒ Ⓓ   5 Ⓐ Ⓑ Ⓒ Ⓓ   8 Ⓐ Ⓑ Ⓒ Ⓓ

3 Ⓐ Ⓑ Ⓒ Ⓓ   6 Ⓐ Ⓑ Ⓒ Ⓓ   9 Ⓐ Ⓑ Ⓒ Ⓓ

Cardboard Folding—Test III

1 Ⓐ Ⓑ Ⓒ Ⓓ   4 Ⓐ Ⓑ Ⓒ Ⓓ   7 Ⓐ Ⓑ Ⓒ Ⓓ

2 Ⓐ Ⓑ Ⓒ Ⓓ   5 Ⓐ Ⓑ Ⓒ Ⓓ   8 Ⓐ Ⓑ Ⓒ Ⓓ

3 Ⓐ Ⓑ Ⓒ Ⓓ   6 Ⓐ Ⓑ Ⓒ Ⓓ   9 Ⓐ Ⓑ Ⓒ Ⓓ

Box Unfolding—Test IV

1 Ⓐ Ⓑ Ⓒ Ⓓ   3 Ⓐ Ⓑ Ⓒ Ⓓ   5 Ⓐ Ⓑ Ⓒ Ⓓ   7 Ⓐ Ⓑ Ⓒ Ⓓ

2 Ⓐ Ⓑ Ⓒ Ⓓ   4 Ⓐ Ⓑ Ⓒ Ⓓ   6 Ⓐ Ⓑ Ⓒ Ⓓ   8 Ⓐ Ⓑ Ⓒ Ⓓ

Box Unfolding—Test V

1 Ⓐ Ⓑ Ⓒ Ⓓ   3 Ⓐ Ⓑ Ⓒ Ⓓ   5 Ⓐ Ⓑ Ⓒ Ⓓ   7 Ⓐ Ⓑ Ⓒ Ⓓ

2 Ⓐ Ⓑ Ⓒ Ⓓ   4 Ⓐ Ⓑ Ⓒ Ⓓ   6 Ⓐ Ⓑ Ⓒ Ⓓ   8 Ⓐ Ⓑ Ⓒ Ⓓ

Box Unfolding—Test VI

1 Ⓐ Ⓑ Ⓒ Ⓓ   3 Ⓐ Ⓑ Ⓒ Ⓓ   5 Ⓐ Ⓑ Ⓒ Ⓓ   7 Ⓐ Ⓑ Ⓒ Ⓓ

2 Ⓐ Ⓑ Ⓒ Ⓓ   4 Ⓐ Ⓑ Ⓒ Ⓓ   6 Ⓐ Ⓑ Ⓒ Ⓓ   8 Ⓐ Ⓑ Ⓒ Ⓓ

FIGURE TURNING

Solid Figure Turning—Test I

1 Ⓐ Ⓑ Ⓒ Ⓓ   6 Ⓐ Ⓑ Ⓒ Ⓓ   11 Ⓐ Ⓑ Ⓒ Ⓓ   16 Ⓐ Ⓑ Ⓒ Ⓓ   21 Ⓐ Ⓑ Ⓒ Ⓓ

2 Ⓐ Ⓑ Ⓒ Ⓓ   7 Ⓐ Ⓑ Ⓒ Ⓓ   12 Ⓐ Ⓑ Ⓒ Ⓓ   17 Ⓐ Ⓑ Ⓒ Ⓓ   22 Ⓐ Ⓑ Ⓒ Ⓓ

3 Ⓐ Ⓑ Ⓒ Ⓓ   8 Ⓐ Ⓑ Ⓒ Ⓓ   13 Ⓐ Ⓑ Ⓒ Ⓓ   18 Ⓐ Ⓑ Ⓒ Ⓓ

4 Ⓐ Ⓑ Ⓒ Ⓓ   9 Ⓐ Ⓑ Ⓒ Ⓓ   14 Ⓐ Ⓑ Ⓒ Ⓓ   19 Ⓐ Ⓑ Ⓒ Ⓓ

5 Ⓐ Ⓑ Ⓒ Ⓓ   10 Ⓐ Ⓑ Ⓒ Ⓓ   15 Ⓐ Ⓑ Ⓒ Ⓓ   20 Ⓐ Ⓑ Ⓒ Ⓓ

Cube Turning—Test II

1 Ⓐ Ⓑ Ⓒ Ⓓ   7 Ⓐ Ⓑ Ⓒ Ⓓ   13 Ⓐ Ⓑ Ⓒ Ⓓ

2 Ⓐ Ⓑ Ⓒ Ⓓ   8 Ⓐ Ⓑ Ⓒ Ⓓ   14 Ⓐ Ⓑ Ⓒ Ⓓ

3 Ⓐ Ⓑ Ⓒ Ⓓ   9 Ⓐ Ⓑ Ⓒ Ⓓ   15 Ⓐ Ⓑ Ⓒ Ⓓ

4 Ⓐ Ⓑ Ⓒ Ⓓ   10 Ⓐ Ⓑ Ⓒ Ⓓ   16 Ⓐ Ⓑ Ⓒ Ⓓ

5 Ⓐ Ⓑ Ⓒ Ⓓ   11 Ⓐ Ⓑ Ⓒ Ⓓ   17 Ⓐ Ⓑ Ⓒ Ⓓ

6 Ⓐ Ⓑ Ⓒ Ⓓ   12 Ⓐ Ⓑ Ⓒ Ⓓ   18 Ⓐ Ⓑ Ⓒ Ⓓ

PERCEPTUAL ABILITY

Similarities and Differences Between Objects—Test I

1 Ⓐ Ⓑ Ⓒ Ⓓ Ⓔ   6 Ⓐ Ⓑ Ⓒ Ⓓ Ⓔ   11 Ⓐ Ⓑ Ⓒ Ⓓ Ⓔ

2 Ⓐ Ⓑ Ⓒ Ⓓ Ⓔ   7 Ⓐ Ⓑ Ⓒ Ⓓ Ⓔ   12 Ⓐ Ⓑ Ⓒ Ⓓ Ⓔ

3 Ⓐ Ⓑ Ⓒ Ⓓ Ⓔ   8 Ⓐ Ⓑ Ⓒ Ⓓ Ⓔ   13 Ⓐ Ⓑ Ⓒ Ⓓ Ⓔ

4 Ⓐ Ⓑ Ⓒ Ⓓ Ⓔ   9 Ⓐ Ⓑ Ⓒ Ⓓ Ⓔ   14 Ⓐ Ⓑ Ⓒ Ⓓ Ⓔ

5 Ⓐ Ⓑ Ⓒ Ⓓ Ⓔ   10 Ⓐ Ⓑ Ⓒ Ⓓ Ⓔ   15 Ⓐ Ⓑ Ⓒ Ⓓ Ⓔ

Line Ranking—Test II

1 ① ② ③ ④   12 ① ② ③ ④

2 ① ② ③ ④   13 ① ② ③ ④

3 ① ② ③ ④   14 ① ② ③ ④

4 ① ② ③ ④   15 ① ② ③ ④

5 ① ② ③ ④   16 ① ② ③ ④

6 ① ② ③ ④   17 ① ② ③ ④

7 ① ② ③ ④   18 ① ② ③ ④

8 ① ② ③ ④   19 ① ② ③ ④

9 ① ② ③ ④   20 ① ② ③ ④

10 ① ② ③ ④   21 ① ② ③ ④

11 ① ② ③ ④   22 ① ② ③ ④

Angle Comparisons—Test III

1. _____   12. _____

2. _____   13. _____

3. _____   14. _____

4. _____   15. _____

5. _____   16. _____

6. _____   17. _____

7. _____   18. _____

8. _____   19. _____

9. _____   20. _____

10. _____   21. _____

11. _____   22. _____

VISUAL-MOTOR COORDINATION
Letter-Symbol Coding—Test I

1. _____	26. _____	51. _____	76. _____
2. _____	27. _____	52. _____	77. _____
3. _____	28. _____	53. _____	78. _____
4. _____	29. _____	54. _____	79. _____
5. _____	30. _____	55. _____	80. _____
6. _____	31. _____	56. _____	81. _____
7. _____	32. _____	57. _____	82. _____
8. _____	33. _____	58. _____	83. _____
9. _____	34. _____	59. _____	84. _____
10. _____	35. _____	60. _____	85. _____
11. _____	36. _____	61. _____	86. _____
12. _____	37. _____	62. _____	87. _____
13. _____	38. _____	63. _____	88. _____
14. _____	39. _____	64. _____	89. _____
15. _____	40. _____	65. _____	90. _____
16. _____	41. _____	66. _____	91. _____
17. _____	42. _____	67. _____	92. _____
18. _____	43. _____	68. _____	93. _____
19. _____	44. _____	69. _____	94. _____
20. _____	45. _____	70. _____	95. _____
21. _____	46. _____	71. _____	96. _____
22. _____	47. _____	72. _____	97. _____
23. _____	48. _____	73. _____	98. _____
24. _____	49. _____	74. _____	99. _____
25. _____	50. _____	75. _____	100. _____

Counting Crosses and Zeros—Test II

1 Ⓐ Ⓑ Ⓒ Ⓓ Ⓔ	7 Ⓐ Ⓑ Ⓒ Ⓓ Ⓔ	13 Ⓐ Ⓑ Ⓒ Ⓓ Ⓔ	19 Ⓐ Ⓑ Ⓒ Ⓓ Ⓔ
2 Ⓐ Ⓑ Ⓒ Ⓓ Ⓔ	8 Ⓐ Ⓑ Ⓒ Ⓓ Ⓔ	14 Ⓐ Ⓑ Ⓒ Ⓓ Ⓔ	20 Ⓐ Ⓑ Ⓒ Ⓓ Ⓔ
3 Ⓐ Ⓑ Ⓒ Ⓓ Ⓔ	9 Ⓐ Ⓑ Ⓒ Ⓓ Ⓔ	15 Ⓐ Ⓑ Ⓒ Ⓓ Ⓔ	21 Ⓐ Ⓑ Ⓒ Ⓓ Ⓔ
4 Ⓐ Ⓑ Ⓒ Ⓓ Ⓔ	10 Ⓐ Ⓑ Ⓒ Ⓓ Ⓔ	16 Ⓐ Ⓑ Ⓒ Ⓓ Ⓔ	22 Ⓐ Ⓑ Ⓒ Ⓓ Ⓔ
5 Ⓐ Ⓑ Ⓒ Ⓓ Ⓔ	11 Ⓐ Ⓑ Ⓒ Ⓓ Ⓔ	17 Ⓐ Ⓑ Ⓒ Ⓓ Ⓔ	23 Ⓐ Ⓑ Ⓒ Ⓓ Ⓔ
6 Ⓐ Ⓑ Ⓒ Ⓓ Ⓔ	12 Ⓐ Ⓑ Ⓒ Ⓓ Ⓔ	18 Ⓐ Ⓑ Ⓒ Ⓓ Ⓔ	24 Ⓐ Ⓑ Ⓒ Ⓓ Ⓔ

Examining Objects—Test III

1 Ⓐ Ⓑ Ⓒ Ⓓ Ⓔ	10 Ⓐ Ⓑ Ⓒ Ⓓ Ⓔ	19 Ⓐ Ⓑ Ⓒ Ⓓ Ⓔ	28 Ⓐ Ⓑ Ⓒ Ⓓ Ⓔ
2 Ⓐ Ⓑ Ⓒ Ⓓ Ⓔ	11 Ⓐ Ⓑ Ⓒ Ⓓ Ⓔ	20 Ⓐ Ⓑ Ⓒ Ⓓ Ⓔ	29 Ⓐ Ⓑ Ⓒ Ⓓ Ⓔ
3 Ⓐ Ⓑ Ⓒ Ⓓ Ⓔ	12 Ⓐ Ⓑ Ⓒ Ⓓ Ⓔ	21 Ⓐ Ⓑ Ⓒ Ⓓ Ⓔ	30 Ⓐ Ⓑ Ⓒ Ⓓ Ⓔ
4 Ⓐ Ⓑ Ⓒ Ⓓ Ⓔ	13 Ⓐ Ⓑ Ⓒ Ⓓ Ⓔ	22 Ⓐ Ⓑ Ⓒ Ⓓ Ⓔ	31 Ⓐ Ⓑ Ⓒ Ⓓ Ⓔ
5 Ⓐ Ⓑ Ⓒ Ⓓ Ⓔ	14 Ⓐ Ⓑ Ⓒ Ⓓ Ⓔ	23 Ⓐ Ⓑ Ⓒ Ⓓ Ⓔ	32 Ⓐ Ⓑ Ⓒ Ⓓ Ⓔ
6 Ⓐ Ⓑ Ⓒ Ⓓ Ⓔ	15 Ⓐ Ⓑ Ⓒ Ⓓ Ⓔ	24 Ⓐ Ⓑ Ⓒ Ⓓ Ⓔ	33 Ⓐ Ⓑ Ⓒ Ⓓ Ⓔ
7 Ⓐ Ⓑ Ⓒ Ⓓ Ⓔ	16 Ⓐ Ⓑ Ⓒ Ⓓ Ⓔ	25 Ⓐ Ⓑ Ⓒ Ⓓ Ⓔ	34 Ⓐ Ⓑ Ⓒ Ⓓ Ⓔ
8 Ⓐ Ⓑ Ⓒ Ⓓ Ⓔ	17 Ⓐ Ⓑ Ⓒ Ⓓ Ⓔ	26 Ⓐ Ⓑ Ⓒ Ⓓ Ⓔ	35 Ⓐ Ⓑ Ⓒ Ⓓ Ⓔ
9 Ⓐ Ⓑ Ⓒ Ⓓ Ⓔ	18 Ⓐ Ⓑ Ⓒ Ⓓ Ⓔ	27 Ⓐ Ⓑ Ⓒ Ⓓ Ⓔ	36 Ⓐ Ⓑ Ⓒ Ⓓ Ⓔ

Mazes and Pursuits—Test IV

1. _____	11. _____	21. _____	31. _____	41. _____
2. _____	12. _____	22. _____	32. _____	42. _____
3. _____	13. _____	23. _____	33. _____	43. _____
4. _____	14. _____	24. _____	34. _____	44. _____
5. _____	15. _____	25. _____	35. _____	45. _____
6. _____	16. _____	26. _____	36. _____	46. _____
7. _____	17. _____	27. _____	37. _____	47. _____
8. _____	18. _____	28. _____	38. _____	48. _____
9. _____	19. _____	29. _____	39. _____	49. _____
10. _____	20. _____	30. _____	40. _____	50. _____

MAP SKILLS

Map Reading Test

1 Ⓐ Ⓑ Ⓒ Ⓓ	4 Ⓐ Ⓑ Ⓒ Ⓓ	7 Ⓐ Ⓑ Ⓒ Ⓓ	10 Ⓐ Ⓑ Ⓒ Ⓓ	13 Ⓐ Ⓑ Ⓒ Ⓓ
2 Ⓐ Ⓑ Ⓒ Ⓓ	5 Ⓐ Ⓑ Ⓒ Ⓓ	8 Ⓐ Ⓑ Ⓒ Ⓓ	11 Ⓐ Ⓑ Ⓒ Ⓓ	14 Ⓐ Ⓑ Ⓒ Ⓓ
3 Ⓐ Ⓑ Ⓒ Ⓓ	6 Ⓐ Ⓑ Ⓒ Ⓓ	9 Ⓐ Ⓑ Ⓒ Ⓓ	12 Ⓐ Ⓑ Ⓒ Ⓓ	15 Ⓐ Ⓑ Ⓒ Ⓓ

HIDDEN FIGURES

Hidden Figures test your ability to find a simple geometric pattern within a larger, more complex pattern. Each test presents a set of five Simple Patterns labeled A, B, C, D, and E. These simple patterns are followed by a group of more Complex Patterns, which are numbered starting with 1. For each numbered pattern, you are to choose the one lettered pattern that can be found hidden within the more complex form. Try the four sample questions first. Then go on to do the three Hidden Figures Tests that follow. Correct answers to all Hidden Figures Tests are consolidated at the end of this chapter.

FOUR SAMPLE QUESTIONS FOR PRACTICE

STEP ONE: Examine the five Simple Figures below:

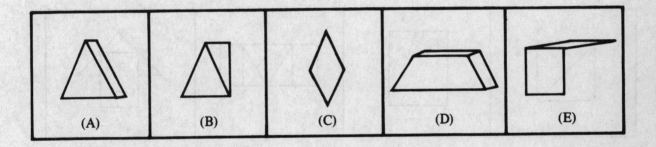

STEP TWO: Try to find the one of the simple figures above that is hidden within each of the more Complex Figures below:

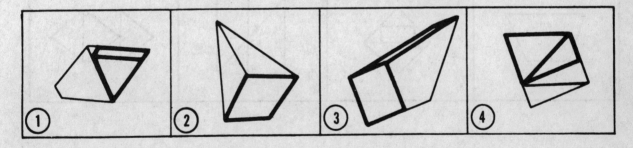

STEP THREE: On your answer sheet, blacken the letter of the Simple Figure found in each numbered Complex Figure. Correct answers to Sample Questions 1 to 4 are shown below and by the darkened lines above.

1 ●ⒷⒸⒹⒺ   2 Ⓐ Ⓑ ●ⒹⒺ   3 ⒶⒷⒸⒹ●   4 Ⓐ ● ⒸⒹⒺ

TEST I. HIDDEN FIGURES

Simple Figures for Questions 1 to 28

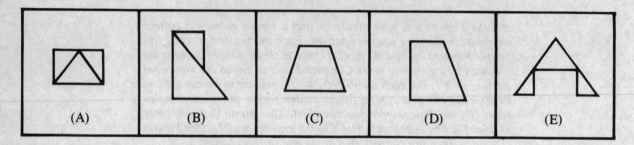

Find the Simple Figures hidden in these more Complex Figures. On your answer sheet, blacken the letter that corresponds to the Simple Figure you find in each Complex Figure below.

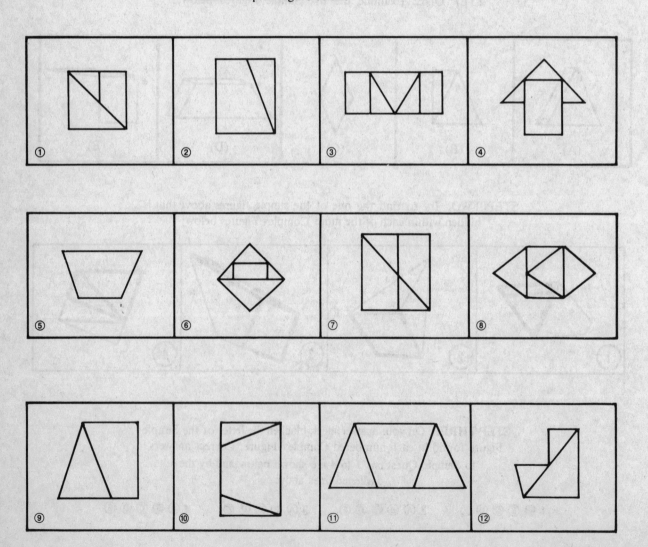

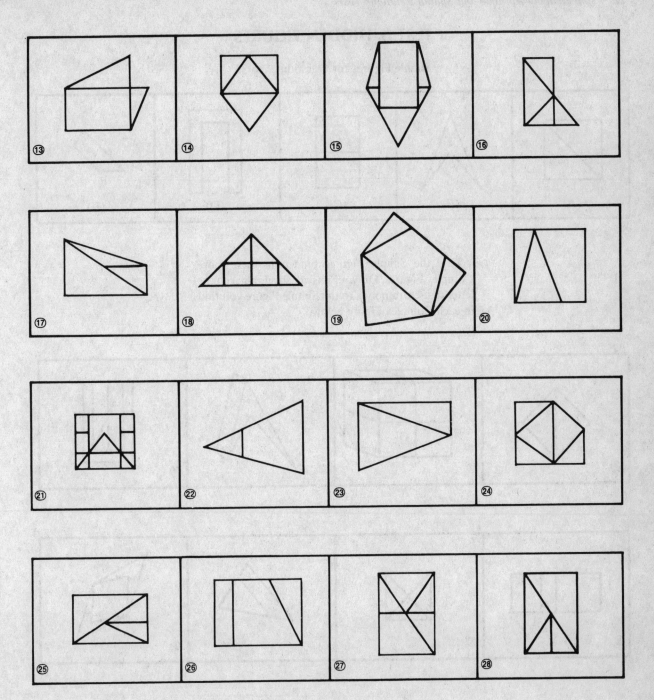

TEST II. HIDDEN FIGURES

Simple Figures for Questions 1 to 32

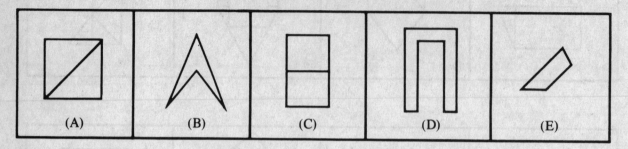

(A)   (B)   (C)   (D)   (E)

Find the Simple Figures hidden in these more Complex Figures. On your answer sheet, blacken the letter that corresponds to the Simple Figure you find in each Complex Figure below.

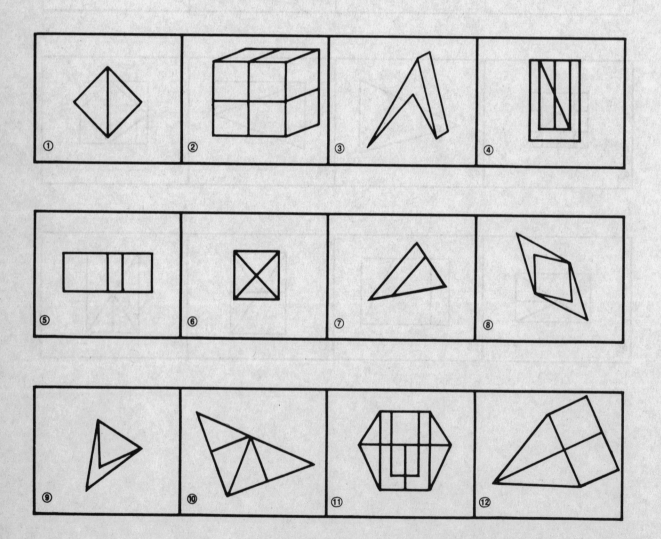

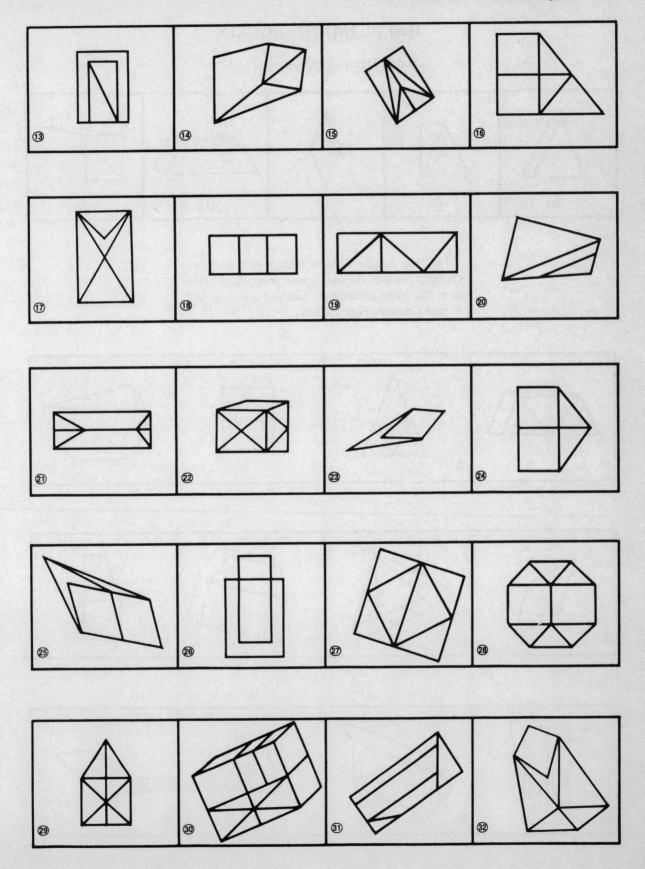

TEST III. HIDDEN FIGURES

Simple Figures for Questions 1 to 32

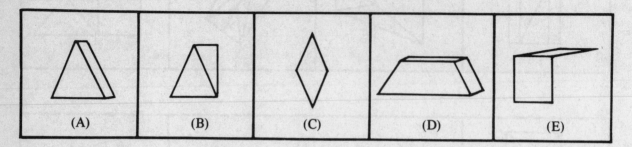

(A) (B) (C) (D) (E)

Find the Simple Figures hidden in these more Complex Figures. On your answer sheet, blacken the letter that corresponds to the Simple Figure you find in each Complex Figure below.

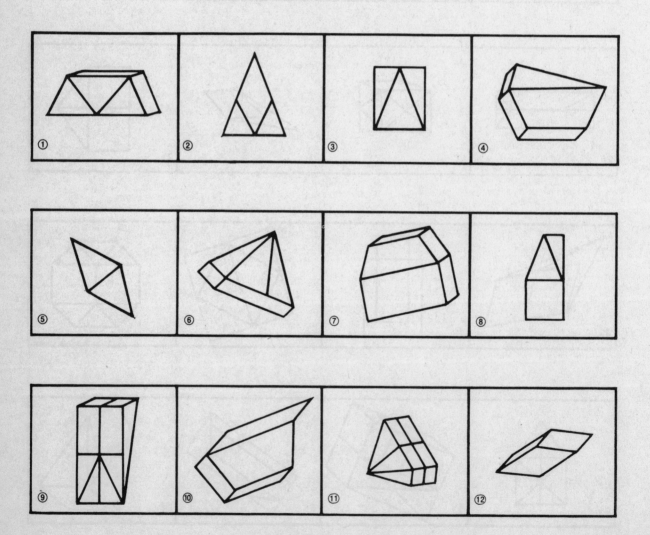

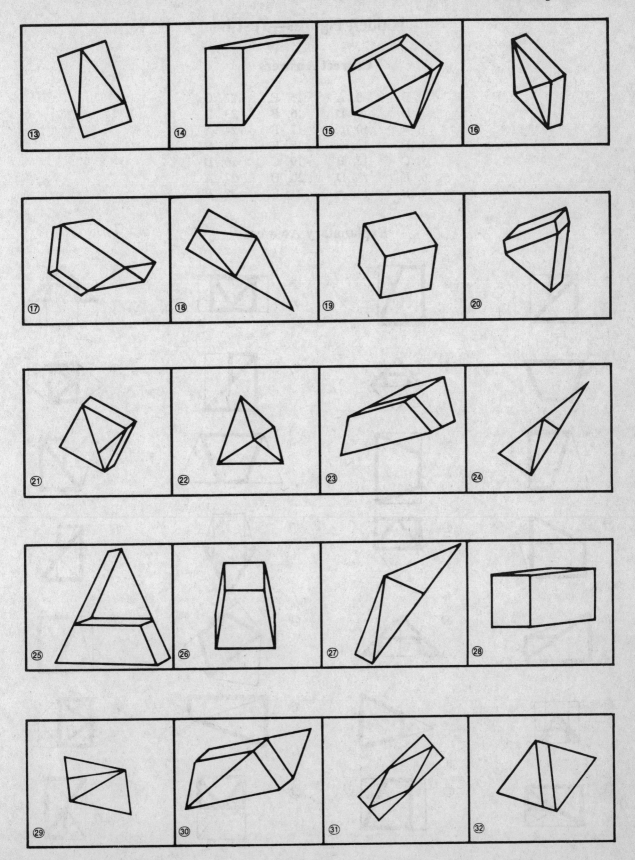

Hidden Figures—Test I

Correct Answers

1. B	8. A	15. E	22. C
2. D	9. D	16. B	23. D
3. A	10. C	17. B	24. A
4. E	11. C	18. E	25. B
5. C	12. B	19. C	26. D
6. E	13. D	20. D	27. A
7. B	14. A	21. E	28. B

Explanatory Answers

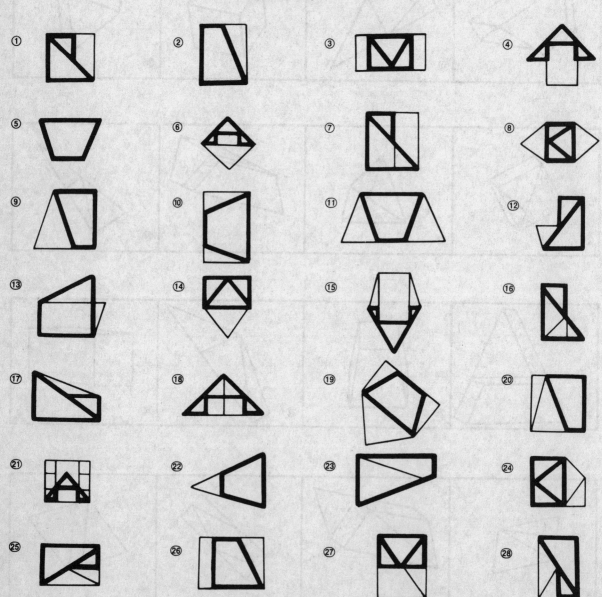

Hidden Figures—Test II

Correct Answers

1. A	9. B	17. B	25. B
2. C	10. A	18. C	26. D
3. B	11. D	19. A	27. A
4. D	12. C	20. E	28. C
5. C	13. D	21. E	29. A
6. A	14. A	22. A	30. C
7. E	15. B	23. B	31. E
8. B	16. C	24. C	32. B

Explanatory Answers

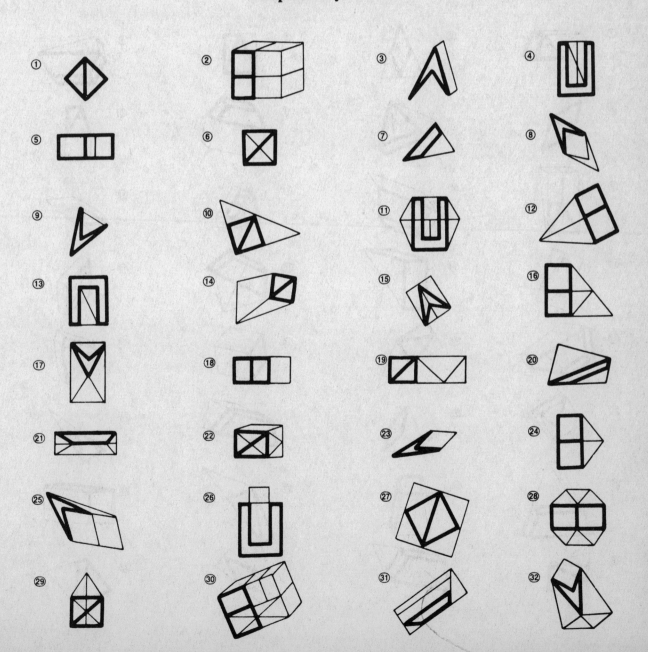

Hidden Figures—Test III

Correct Answers

1. A	9. B	17. D	25. D
2. C	10. D	18. B	26. E
3. B	11. A	19. C	27. B
4. D	12. C	20. A	28. E
5. C	13. B	21. A	29. B
6. A	14. E	22. C	30. D
7. D	15. D	23. D	31. C
8. B	16. C	24. B	32. A

Explanatory Answers

MATCHING PARTS AND FIGURES

There are several types of questions that test your ability to match parts and figures. Although the directions may vary slightly, each question type requires you to visualize the shape or pattern that can result from fitting together a number of cut-up pieces.

Questions 1 and 2 below illustrate one type of Parts and Figures question. In these questions, you are given one numbered figure and a group of five lettered figures. You are to choose the two lettered figures which, when put together, make a figure of the same size and shape as the numbered figure.

Look at question 1. The numbered figure is a circle. Each lettered figure is a part of a circle; however, only alternatives A and E will fit together to make a complete circle of the same size as the given circle.

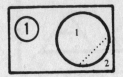 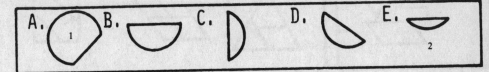

Now try question 2.

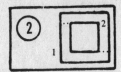 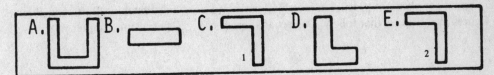

Only alternatives C and E will fit together to form the square within a square illustrated by the given figure.

Question 3 shows another variation on matching parts and figures. In this question you are given a numbered pattern together with four lettered groups of cut-up pattern pieces. Here the task is to determine which group of pieces will fit together to form the numbered pattern. The cut-up pieces may be turned around or turned over to make them fit the given pattern, but there can be no spaces between pieces and no overlapping edges.

The only two pieces that will fit together to form the pattern shown in question 3 are the pieces shown in alternative A.

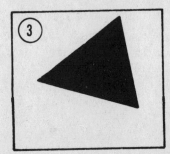 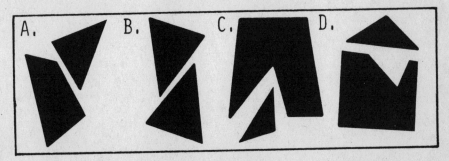

Questions 4 and 5 reverse the procedure used in the first three questions by giving you numbered pattern pieces and asking you to identify the complete pattern that can be made from the pieces shown. As in the other question types, you may have to mentally turn over or turn around the pieces given in order to create the pattern.

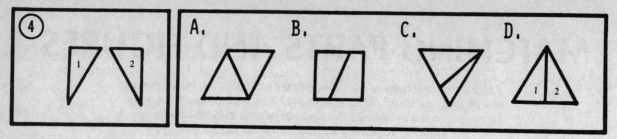

The only arrangement that can be made of the two pieces shown in question 4
is the one illustrated by alternative D.

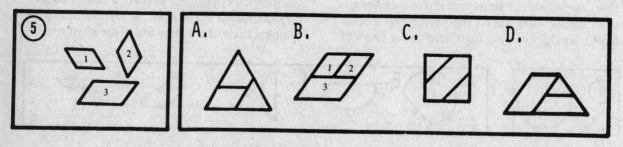

The three pieces shown in question 5 can be arranged only as indicated in alternative B.

You'll get plenty of practice in matching parts and figures in the pages that follow. Correct answers for all Matching Parts and Figures questions are at the end of this chapter.

TEST I. MATCHING PARTS
AND FIGURES

DIRECTIONS: Each of the questions in this test consists of a numbered figure plus a group of five figures lettered A, B, C, D, E. When two of these lettered figures are put together they make up the numbered figure. Choose the letters of the two figures which, when put together, are most nearly the same as the numbered figure.

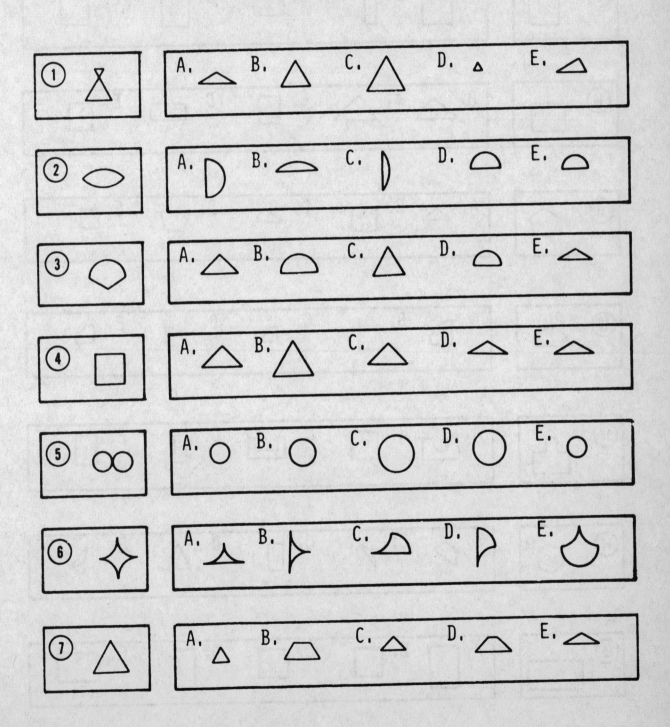

8. A. B. C. D. E.

9. A. B. C. D. E.

10. A. B. C. D. E.

11. A. B. C. D. E.

12. A. B. C. D. E.

13. A. B. C. D. E.

14. A. B. C. D. E.

15. A. B. C. D. E.

TEST II. MATCHING PARTS
AND FIGURES

DIRECTIONS: *Each of the questions in this test consists of a numbered figure plus a group of five figures lettered A, D, C, D, E. When two of these lettered figures are put together they make up the numbered figure. Choose the letters of the two figures which, when put together, are most nearly the same as the numbered figure.*

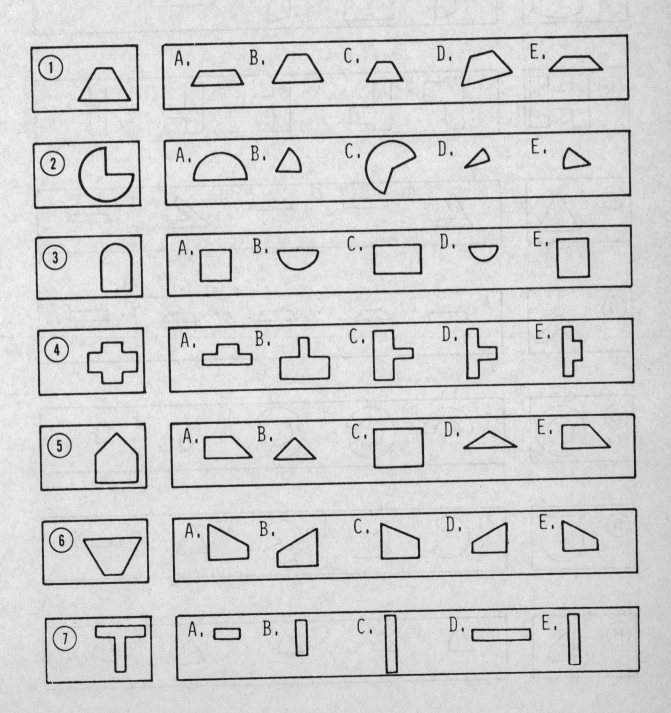

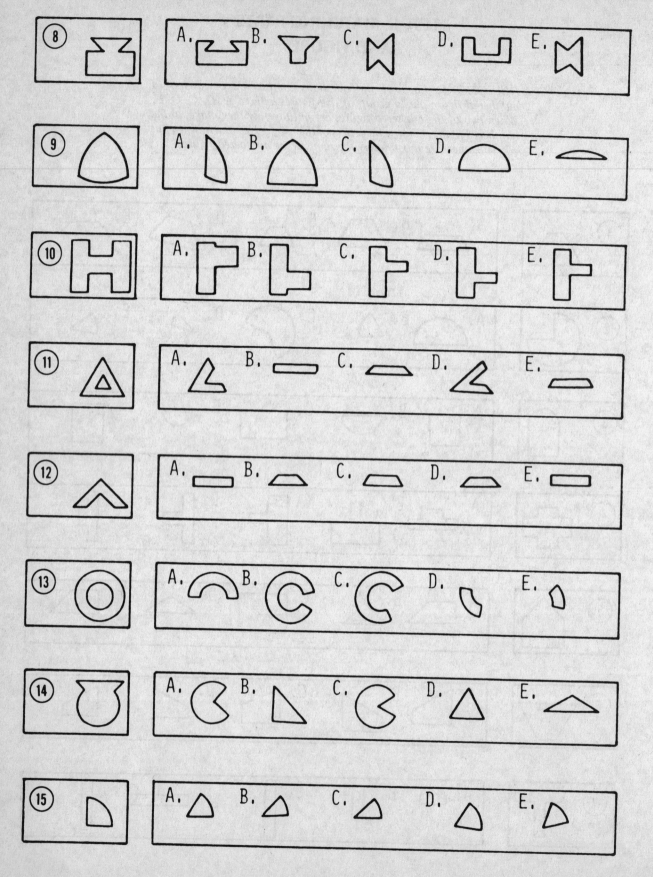

TEST III. MATCHING PARTS
AND FIGURES

DIRECTIONS: Each question in this test consists of a numbered picture showing a single solid pattern and a group of four lettered pictures showing cut-up pattern pieces. For each question, choose the lettered combination of cut-up pieces which, when put together, will make up the numbered pattern shown. The cut-up pieces may be turned around or turned over to make them fit.

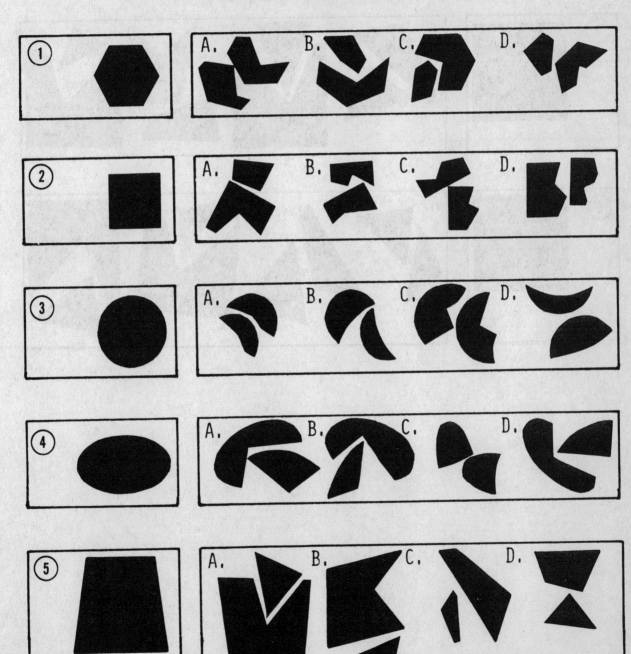

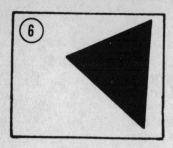
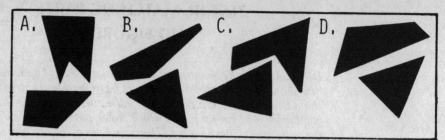

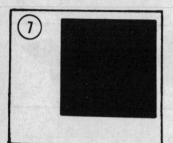
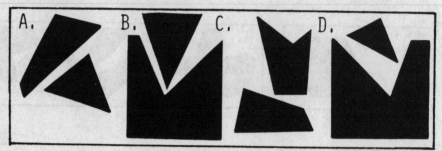

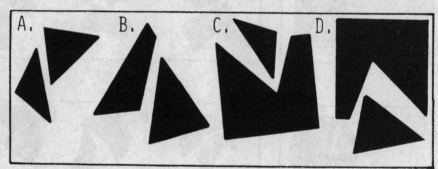

TEST IV. MATCHING PARTS
AND FIGURES

DIRECTIONS: *Each question in this test consists of a numbered picture which shows the parts of an object. To the right of the numbered picture are several objects lettered A, B, C, and D. You are to select that one of the lettered objects which is made up from the numbered parts.*

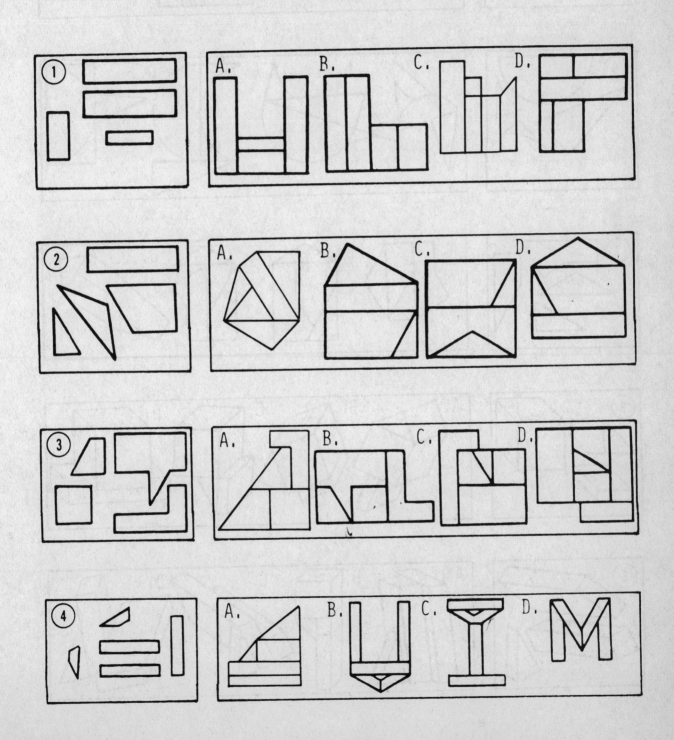

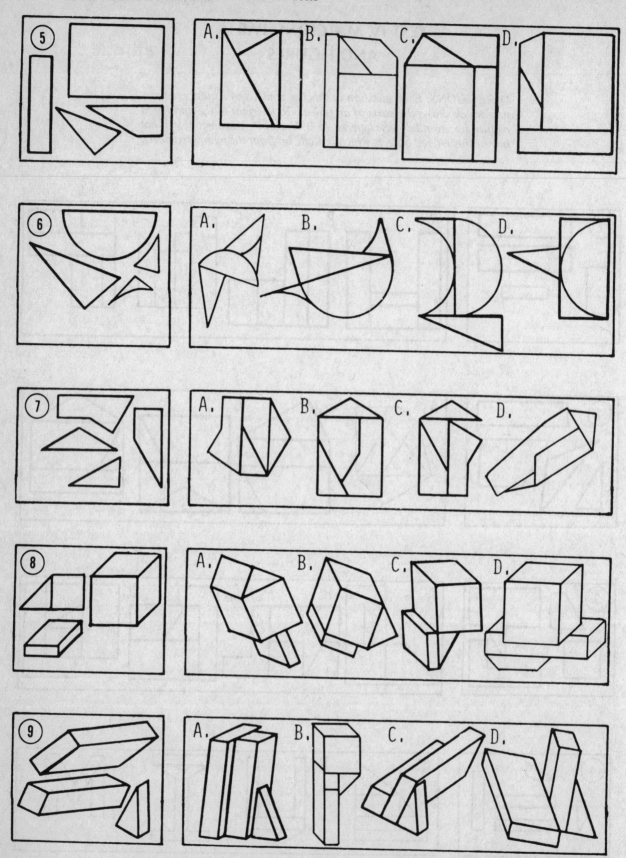

Matching Parts—Test I

Correct Answers

1. B,D	4. A,C	7. A,B	10. A,B	13. A,D
2. B,C	5. A,E	8. B,D	11. A,D	14. A,D
3. B,E	6. A,B	9. B,C	12. D,E	15. B,C

Explanatory Answers

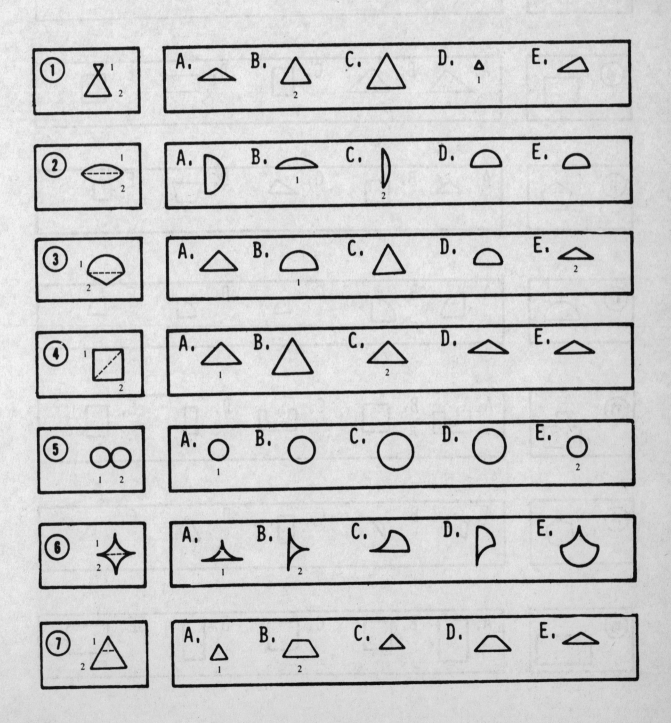

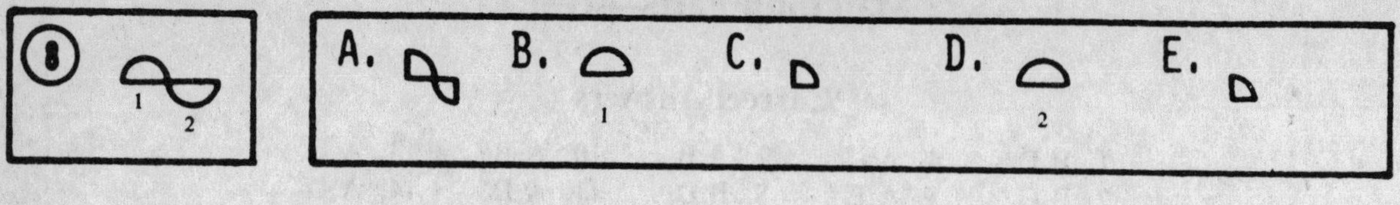
8
1
2
A.
B.
1
C.
D.
2
E.

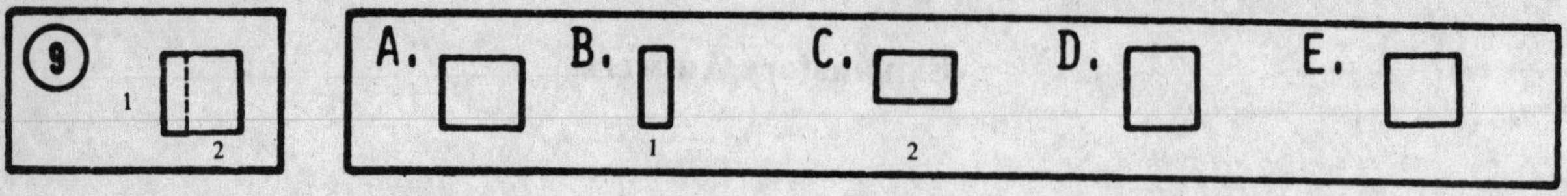
9
1
2
A.
B.
1
C.
2
D.
E.

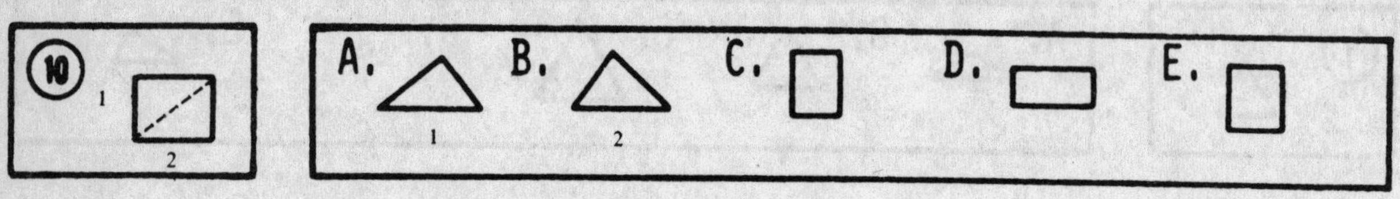
10
1
2
A.
1
B.
2
C.
D.
E.

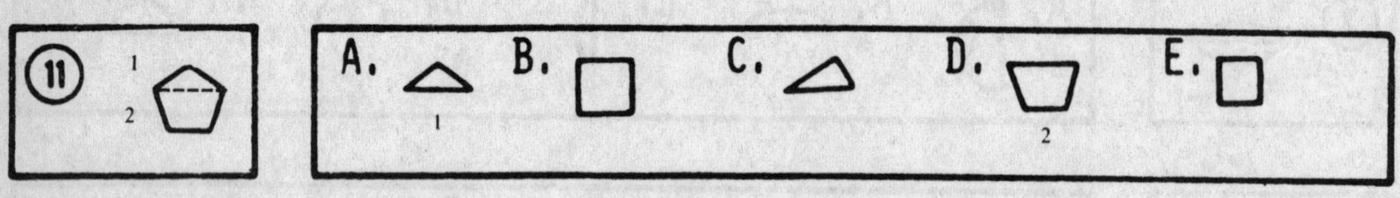
11
1
2
A.
1
B.
C.
D.
2
E.

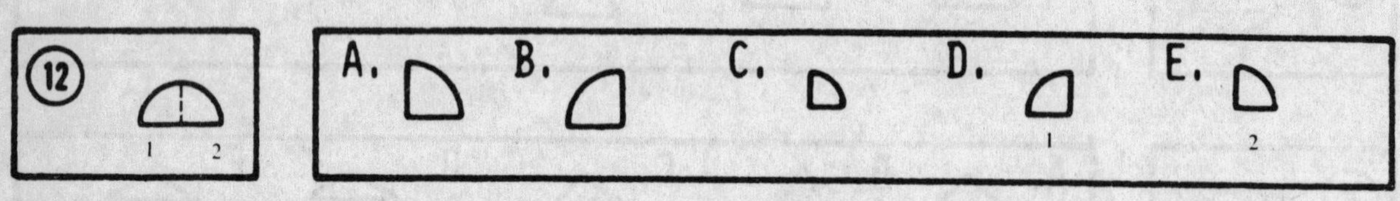
12
1
2
A.
B.
C.
D.
1
E.
2

13
2
1
A.
1
B.
C.
D.
2
E.

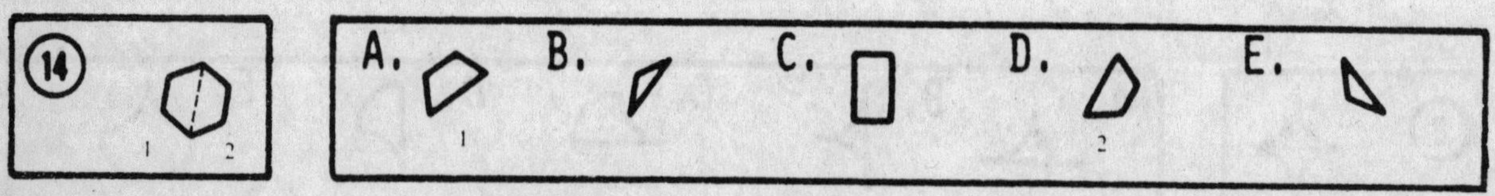
14
1
2
A.
1
B.
C.
D.
2
E.

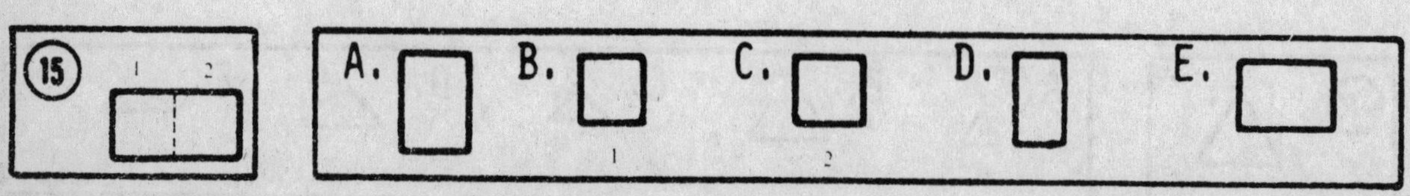
15
1
2
A.
B.
1
C.
2
D.
E.

Matching Parts—Test II

Correct Answers

1. A,C	4. A,E	7. B,E	10. C,E	13. C,D
2. C,E	5. A,E	8. A,C	11. A,E	14. A,B
3. A,D	6. D,E	9. B,E	12. B,D	15. B,C

Explanatory Answers

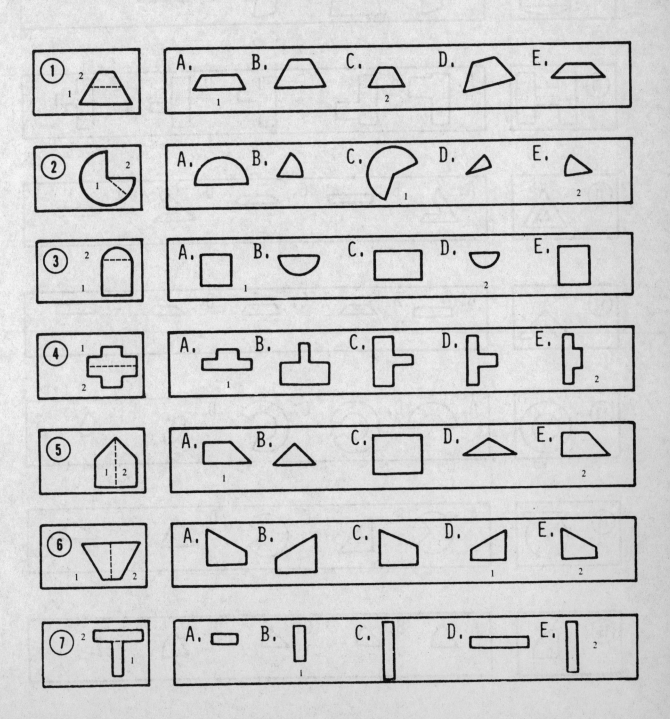

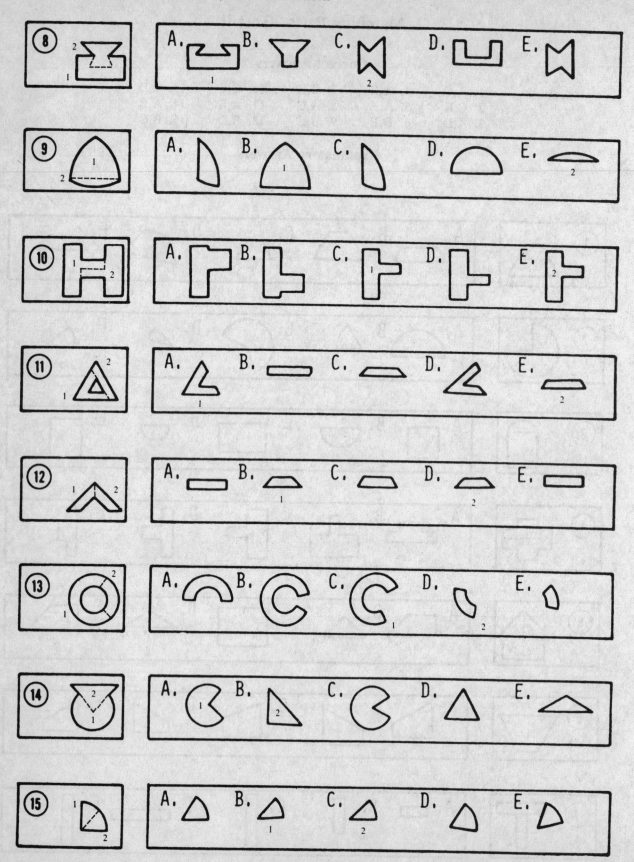

Matching Parts—Test III

Correct Answers

1. A 3. C 5. A 7. B
2. A 4. A 6. D 8. D

Explanatory Answers

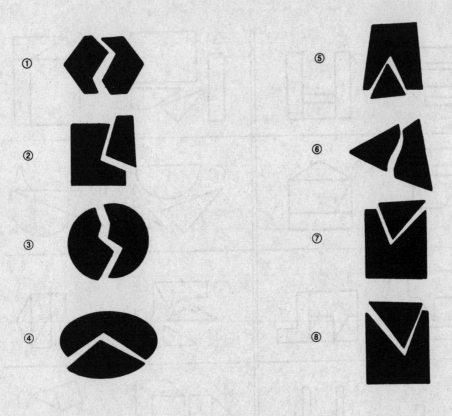

Matching Parts—Test IV

Correct Answers

1. A	4. B	7. A
2. D	5. C	8. A
3. B	6. B	9. C

Explanatory Answers

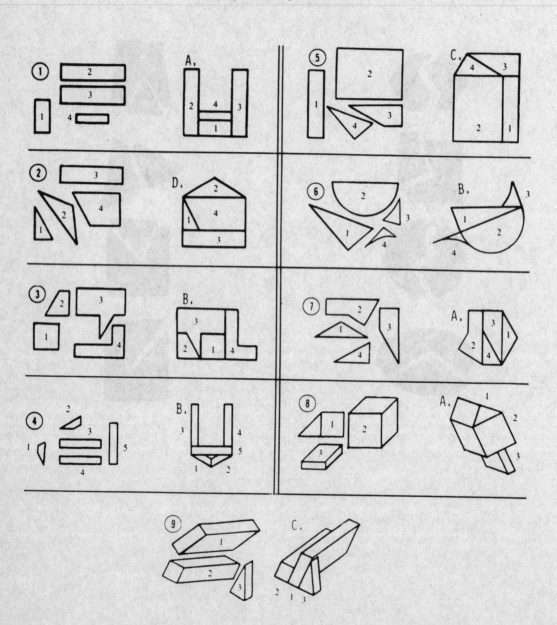

SPATIAL VIEWS

Spatial Views questions test yet another aspect of your ability to comprehend static objects. In View questions, you are given a numbered picture showing the top, front, and side representations of a three-dimensional object. Dashed lines indicate folds. Next to each numbered picture are four lettered drawings. You are to select the one of the lettered drawings that would have the top, front, and side representations shown in the numbered picture.

A SAMPLE QUESTION EXPLAINED

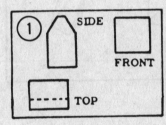 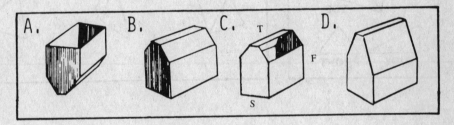

The first frame of question 1 shows the top, side, and front representations of one of the objects labeled A, B, C, and D. At first glance you can eliminate alternative D since the side view is taller and thinner than the side representation shown in the first frame. Alternatives A and B can also be eliminated because they offer front representations which are too long and narrow for the given front view. Alternative C is the only one of the four figures that could have the top, side, and front representations shown in the numbered picture.

Correct Answers to all Spatial Views questions are provided at the end of this chapter.

TEST I. SPATIAL VIEWS

DIRECTIONS: Each question in this test consists of a numbered picture showing the top, front, and side representations of a three-dimensional object. Dashed lines indicate folds. To the right of the numbered representations are four pictures, lettered A, B, C, and D. You are to select one of the drawings that would have the top, front, and side representations shown in the numbered picture. Blacken in your answer sheet accordingly.

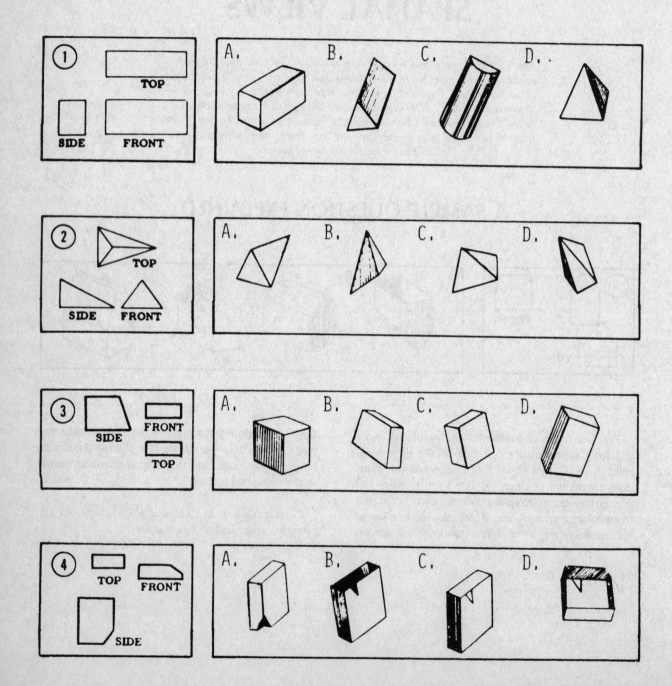

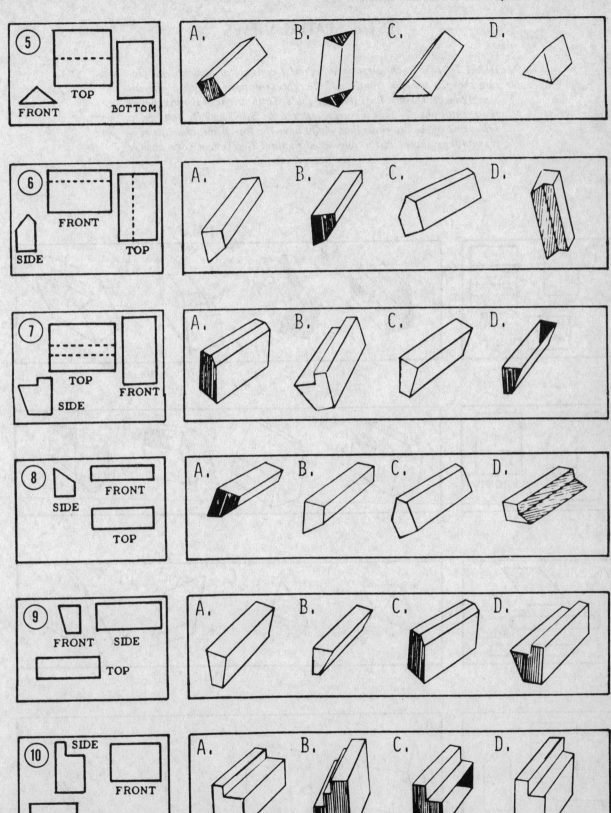

TEST II. SPATIAL VIEWS

DIRECTIONS: Each question in this test consists of a numbered picture showing the top, front, and side representations of a three-dimensional object. Dashed lines indicate folds. To the right of the numbered representations are four pictures, lettered A, B, C, and D. You are to select one of the drawings that would have the top, front, and side representations shown in the numbered picture. Blacken in your answer sheet accordingly.

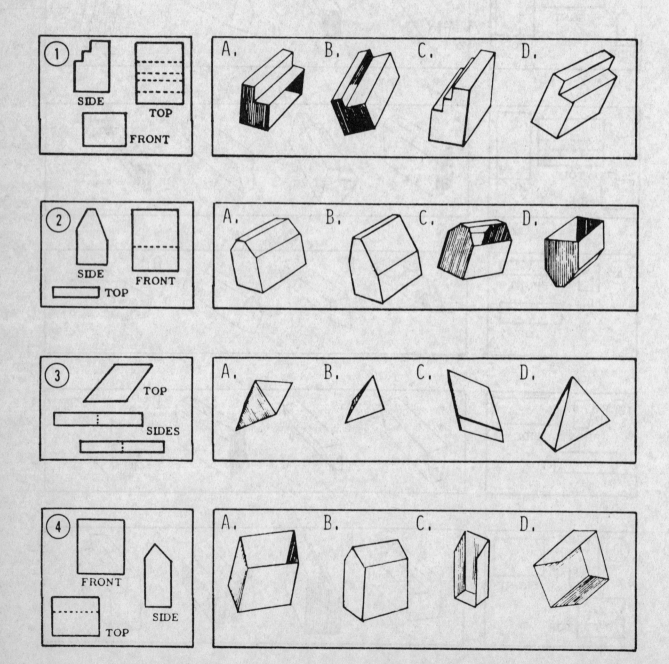

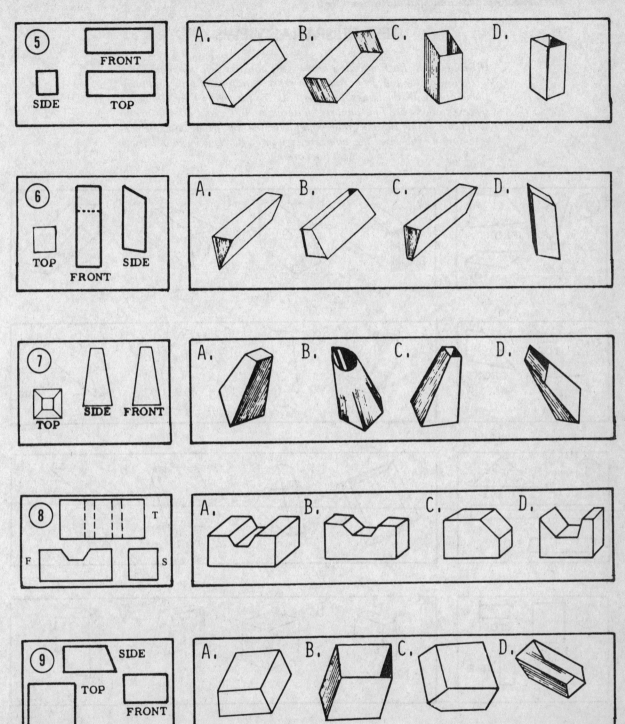

TEST III. SPATIAL VIEWS

DIRECTIONS: Each question in this test consists of a numbered picture showing the top, front, and side representations of a three-dimensional object. Dashed lines indicate folds. To the right of the numbered representations are four pictures, lettered A, B, C, and D. You are to select one of the drawings that would have the top, front, and side representations shown in the numbered picture. Blacken in your answer sheet accordingly.

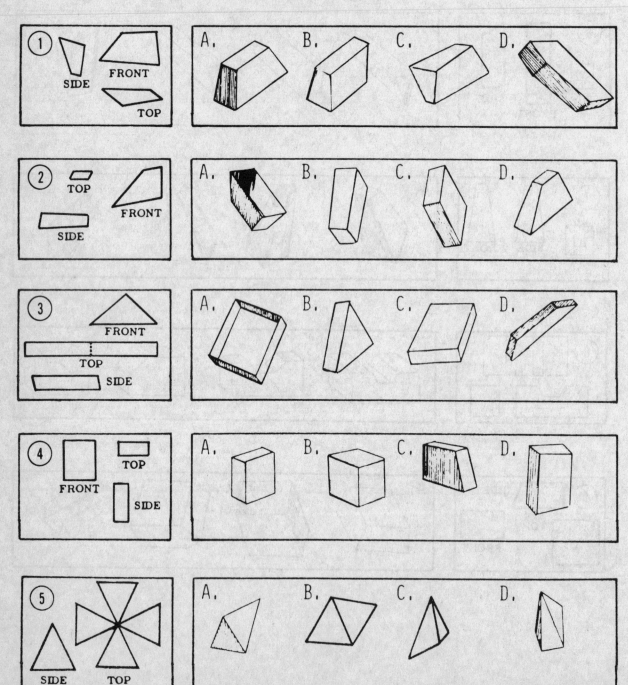

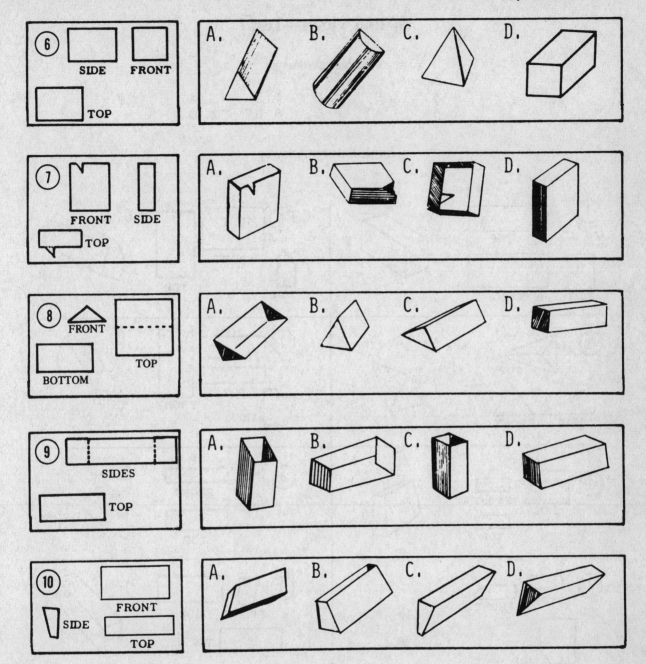

Spatial Views—Test I

Correct Answers

1. A	3. B	5. C	7. B	9. A
2. A	4. A	6. C	8. B	10. D

Explanatory Answers

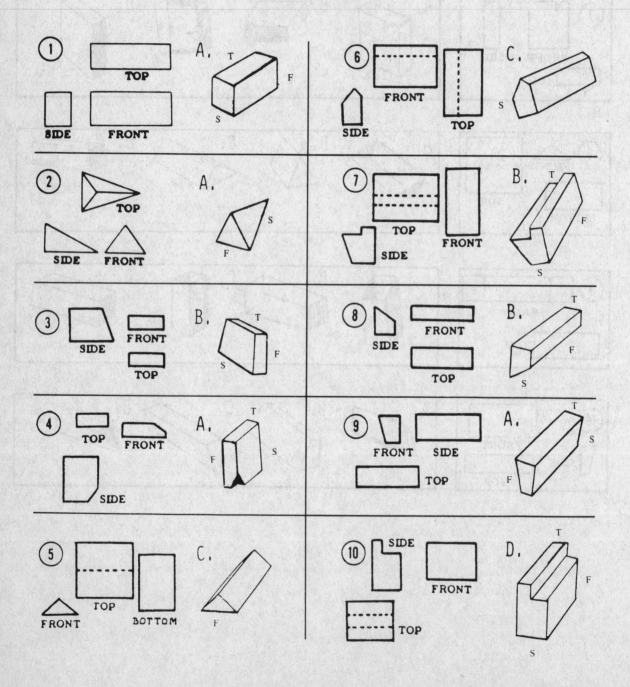

Spatial Views—Test II

Correct Answers

1. C	4. B	7. C
2. B	5. A	8. B
3. C	6. D	9. A

Explanatory Answers

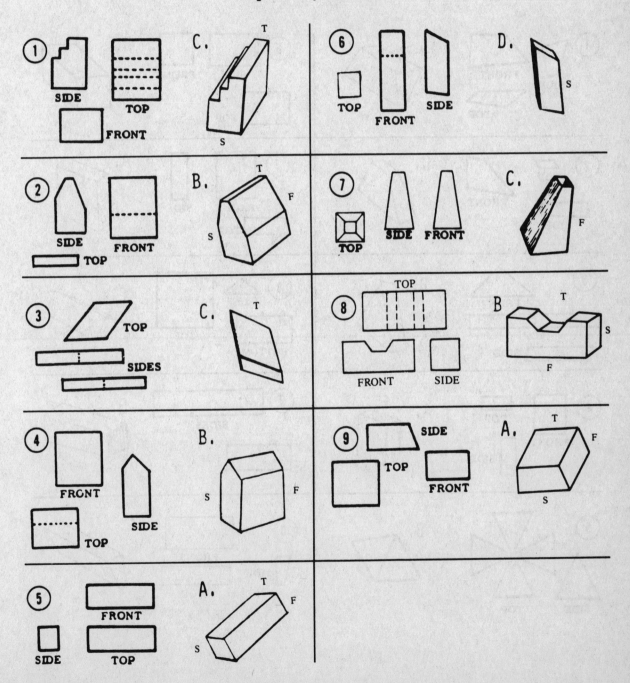

Spatial Views—Test III

Correct Answers

1. C	3. B	5. B	7. A	9. B
2. D	4. A	6. D	8. C	10. C

Explanatory Answers

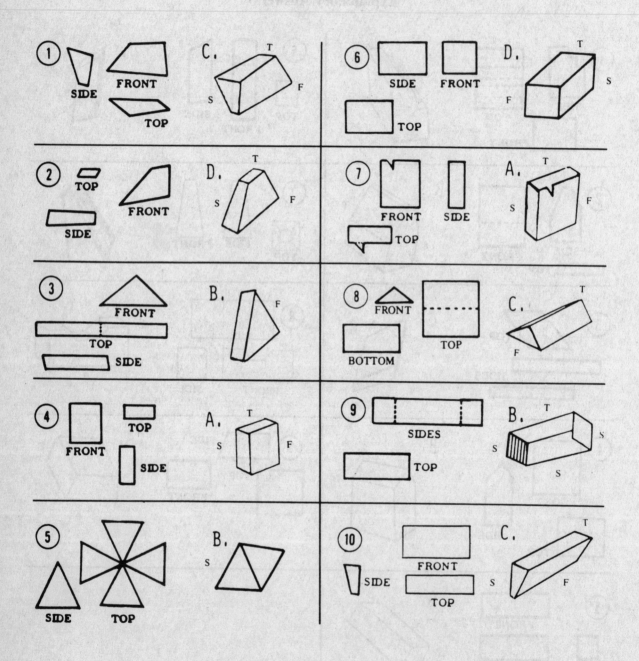

CUBE COUNTING

There are two different question types that measure your ability to count the cubes in a three-dimensional arrangement. In one type you are instructed to count every cube in the arrangement, whether visible or hidden. In the other type of cube counting test, you are to count the number of cubes that touch a particular cube in the group.

Look at sample question 1 below. How many cubes are contained in the entire arrangement?

You should have counted 17 cubes in question 1. There are 4 stacks of 2 cubes each on the left and 4 stacks of 2 cubes each on the right with a single cube connecting the two groups.

A word of caution: Examine each diagram critically, counting the blocks one by one. Even if a figure appears to have four identical parts, do not make the mistake of simply counting the number of blocks in one part and multiplying by 4. You may be counting one row of blocks twice or leaving out a partially hidden column.

Sample question 2 below illustrates another type of cube counting test. For this question, you are to count the number of cubes that touch the cube to which the arrow points.

In touching cube questions, you are to assume that all the cubes in each arrangement are exactly the same size and that there are only enough hidden cubes to support the cubes you can see.

A cube is considered to touch the numbered cube if any part, even a corner, touches.

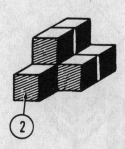

There are 3 cubes that touch the cube numbered 2: the one right behind it which you cannot see, the one on top of that, and the cube that touches the back right corner.

You'll get plenty of practice with both types of cube counting problems on the pages that follow. Correct answers to all cube counting questions may be found at the end of the chapter.

TEST I. CUBE COUNTING

DIRECTIONS: Count the number of blocks in each arrangement. Assume that blocks rest upon blocks immediately beneath them except in the case of arches (as in questions 4, 11, 13 and 17).

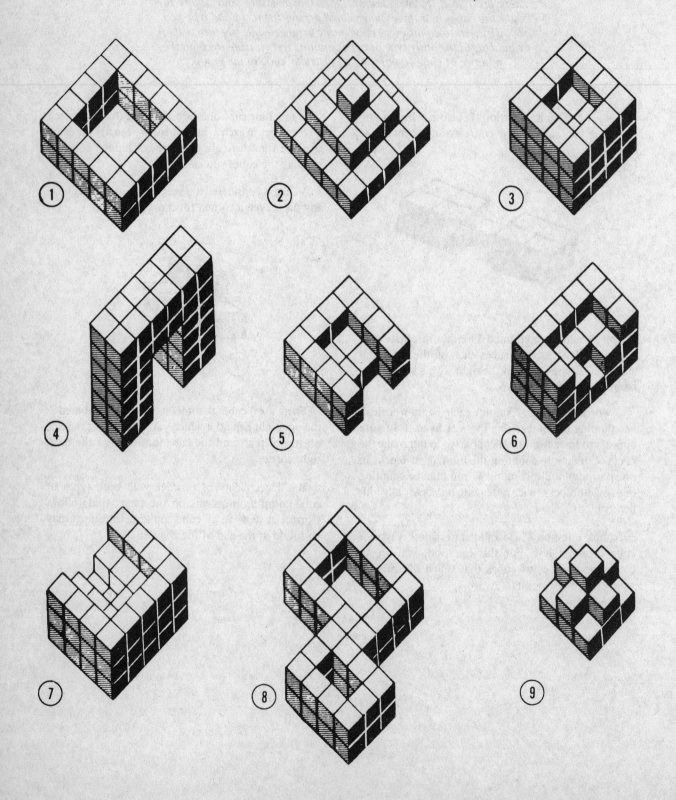

TEST II. CUBE COUNTING

DIRECTIONS: For this test, you are to assume that all the cubes in each arrangement are exactly the same size and that there are only enough hidden cubes to support the cubes you can see. Each question number points to a particular cube. You are to determine how many cubes in the group touch the numbered cube. A cube is considered to touch the numbered cube if any part, even a corner, touches. When you have determined how many cubes touch a given numbered cube, mark your answer sheet as follows:

(A) if the answer is 1, 6 or 11
(B) if the answer is 2, 7 or 12
(C) if the answer is 3, 8 or 13
(D) if the answer is 4, 9 or 14
(E) if the answer is 5, 10 or 15

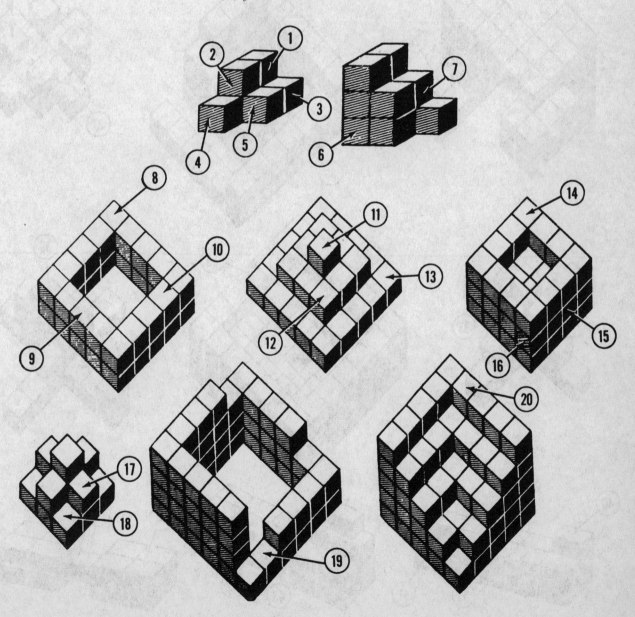

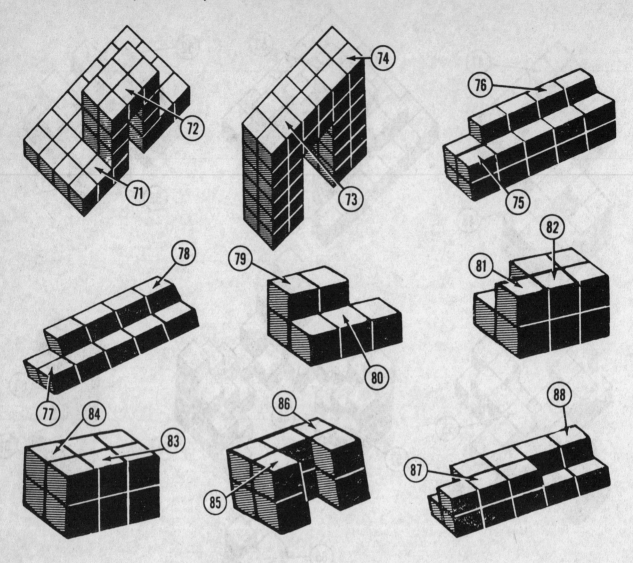

Cube Counting—Test I

Correct Answers

1. 32	6. 38	11. 40	16. 28	21. 70	26. 20
2. 35	7. 49	12. 95	17. 24	22. 58	27. 16
3. 44	8. 48	13. 56	18. 112	23. 47	28. 14
4. 56	9. 14	14. 64	19. 46	24. 105	29. 24
5. 24	10. 59	15. 17	20. 32	25. 17	30. 19

Explanatory Answers

QUESTION	LAYERS						TOTAL
	1	2	3	4	5	6	
1.	16	16					32
2.	25	9	1				35
3.	16	16	12				44
4.	8	8	8	8	12	12	56
5.	14	10					24
6.	15	14	9				38
7.	20	18	11				49
8.	24	24					48
9.	8	5	1				14
10.	20	18	13	8			59
11.	26	4	4	6			40
12.	25	24	21	16	9		95
13.	16	12	6	6	16		56
14.	28	20	12	4			64
15.	13	4					17
16.	13	9	5	1			28
17.	12	6	4	2			24
18.	48	40	24				112
19.	28	18					46
20.	20	12					32
21.	40	21	8	1			70
22.	28	18	9	3			58
23.	19	15	8	4	1		47
24.	49	36	16	4			105
25.	10	7					17
26.	10	9	1				20
27.	10	6					16
28.	10	4					14
29.	10	10	4				24
30.	10	7	2				19

Cube Counting—Test II

Correct Answers

1. E	23. E	45. C	67. A
2. A	24. B	46. A	68. D
3. E	25. A	47. B	69. A
4. C	26. D	48. D	70. D
5. A	27. D	49. D	71. B
6. B	28. B	50. C	72. D
7. E	29. D	51. C	73. A
8. E	30. E	52. A	74. B
9. E	31. E	53. E	75. C
10. B	32. A	54. A	76. C
11. D	33. D	55. A	77. D
12. C	34. E	56. D	78. E
13. D	35. C	57. C	79. E
14. A	36. B	58. C	80. A
15. E	37. A	59. C	81. A
16. E	38. E	60. B	82. E
17. E	39. B	61. D	83. A
18. A	40. D	62. E	84. B
19. E	41. A	63. D	85. E
20. D	42. A	64. A	86. E
21. D	43. D	65. A	87. E
22. C	44. B	66. E	88. E

Explanatory Answers
(Number Touching)

1. 5	23. 5	45. 8	67. 11
2. 6	24. 12	46. 11	68. 4
3. 5	25. 6	47. 7	69. 11
4. 3	26. 9	48. 9	70. 4
5. 6	27. 9	49. 9	71. 7
6. 7	28. 7	50. 8	72. 9
7. 10	29. 9	51. 3	73. 11
8. 5	30. 5	52. 6	74. 7
9. 5	31. 5	53. 5	75. 8
10. 7	32. 6	54. 11	76. 8
11. 9	33. 9	55. 11	77. 4
12. 13	34. 5	56. 9	78. 5
13. 4	35. 8	57. 8	79. 5
14. 6	36. 12	58. 8	80. 6
15. 15	37. 6	59. 13	81. 6
16. 10	38. 5	60. 7	82. 10
17. 10	39. 12	61. 9	83. 11
18. 6	40. 4	62. 15	84. 7
19. 5	41. 6	63. 4	85. 5
20. 9	42. 6	64. 6	86. 5
21. 9	43. 9	65. 11	87. 10
22. 13	44. 7	66. 5	88. 5

PATTERN ANALYSIS

There are two main kinds of questions about boxes. In one the job is to find the box that can be made by folding a piece of cardboard. In the other the task is to determine what the piece of cardboard would look like if a given box were unfolded.

TWO SAMPLE QUESTIONS EXPLAINED

Sample question 1 illustrates a Cardboard Folding question. The numbered frame shows a piece of cardboard that is to be folded along the lines indicated.

You are to determine which of the lettered figures A, B, C or D would be made by folding the cardboard pattern shown.

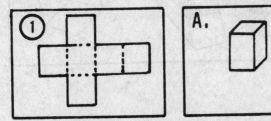 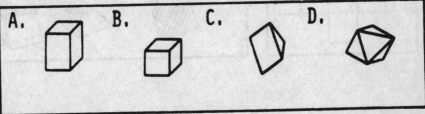

The correct answer to question 1 is B. The pattern shown has 6 equal divisions which, when folded, would form a cube. Neither C nor D is a 6-sided figure and A, which has 4 larger surfaces and 2 smaller ones, could not possibly be formed from the pattern shown.

Now look at Sample Question 2. The numbered frame shows a box with a pattern on each face. If this box were unfolded, it would look like one of the four lettered patterns shown. You are to choose the one lettered pattern that would result from unfolding the numbered box.

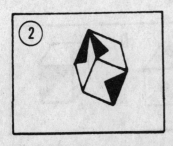 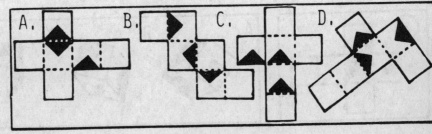

The correct answer to question 2 is D. In alternatives A and C the triangles are incorrectly placed and in alternative B the triangles are on the wrong faces. Only D can be folded to form a cube with each dark triangle in exactly the same position as on the given box.

The practice tests that follow will help you develop your ability to analyze boxes and patterns. Correct answers to all Pattern Analysis questions are consolidated at the end of this chapter.

TEST I. CARDBOARD FOLDING

DIRECTIONS: Each question in this test consists of a numbered picture showing a piece of cardboard that is to be folded. The dotted lines show where folds are to be made. The problem is to choose the lettered picture, A, B, C, or D which would be made by folding the cardboard in the numbered picture. For each question blacken the space on your answer sheet corresponding to the letter of the best answer.

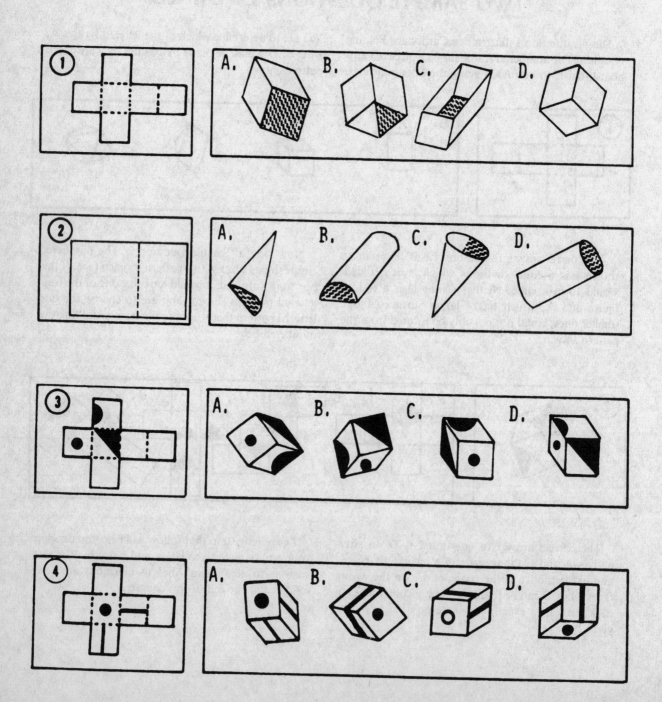

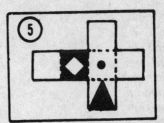
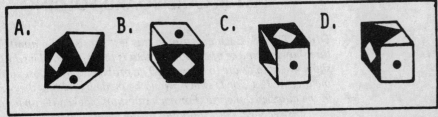

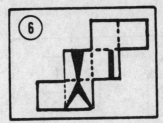
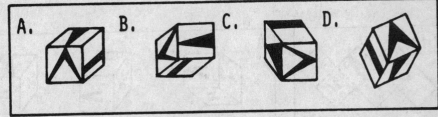

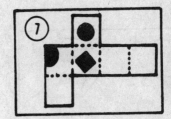
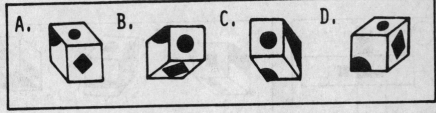

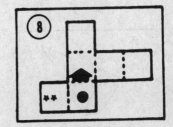
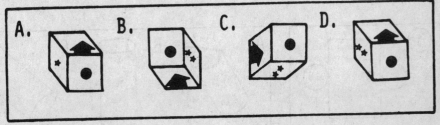

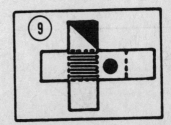

TEST II. CARDBOARD FOLDING

DIRECTIONS: Each question in this test consists of a numbered picture showing a piece of cardboard that is to be folded. The dotted lines show where folds are to be made. The problem is to choose the lettered picture, A, B, C, or D which would be made by folding the cardboard in the numbered picture. For each question blacken the space on your answer sheet corresponding to the letter of the best answer.

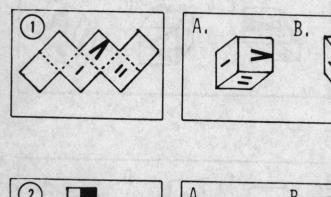

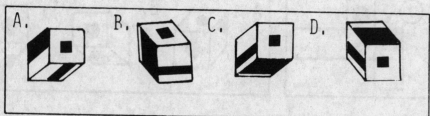

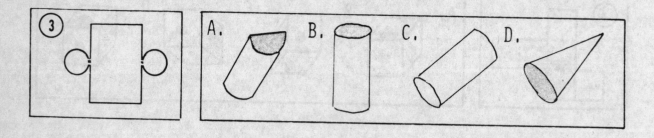

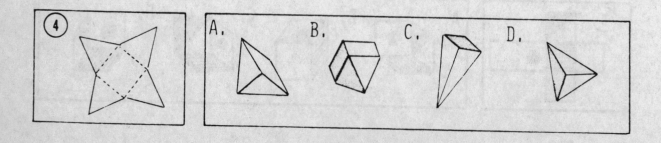

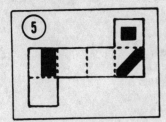

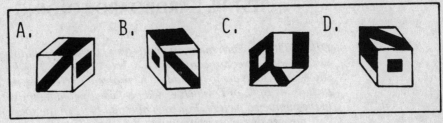

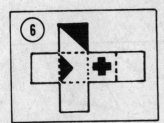

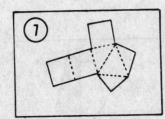

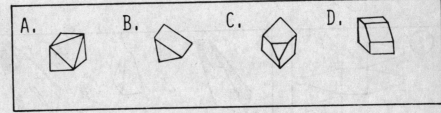

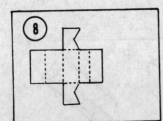

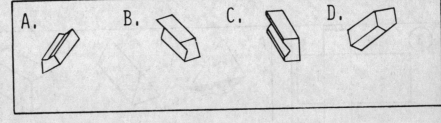

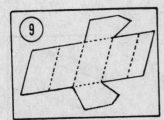

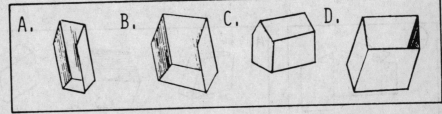

TEST III. CARDBOARD FOLDING

DIRECTIONS: Each question in this tests consists of a numbered picture showing a piece of cardboard that is to be folded. The dotted lines show where folds are to be made. The problem is to choose the lettered picture, A, B, C, or D which would be made by folding the cardboard in the numbered picture. For each question blacken the space on your answer sheet corresponding to the letter of the best answer.

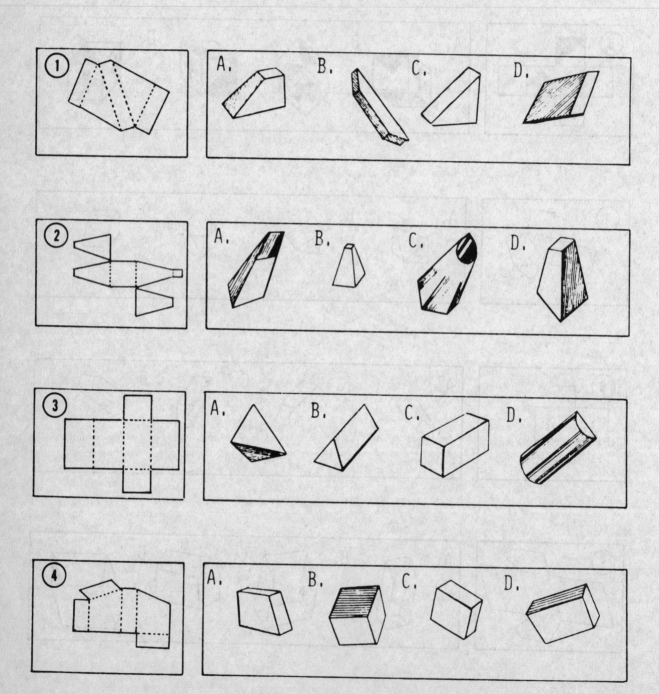

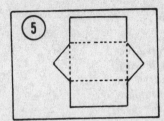

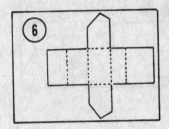
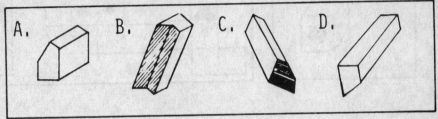

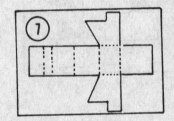
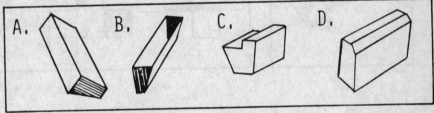

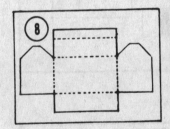
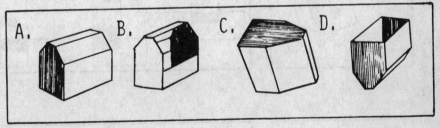

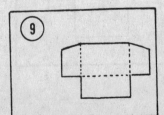
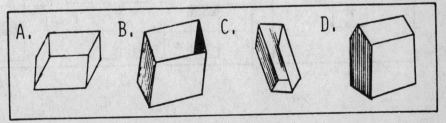

TEST IV. BOX UNFOLDING

DIRECTIONS: *Each question in this test consists of a numbered picture showing a box that is to be unfolded. If the box were unfolded it would look like one of the four cardboard patterns, lettered A, B, C, or D, in the other frames in the row. Choose the cardboard pattern that is unfolded from the lettered picture, and blacken your answer sheet accordingly.*

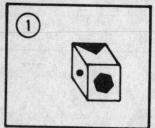
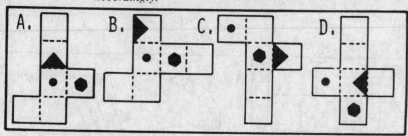

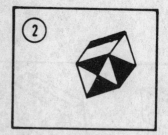
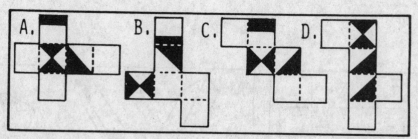

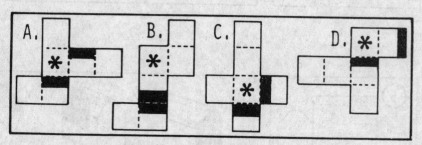

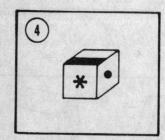
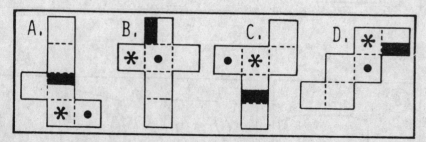

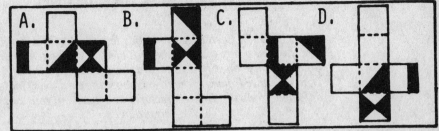

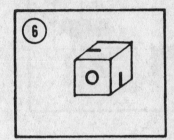
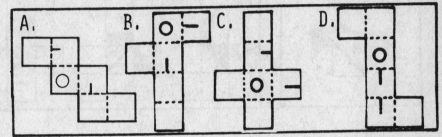

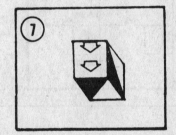
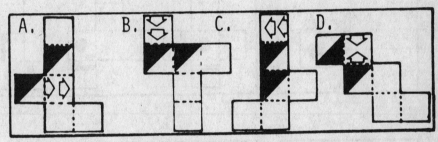

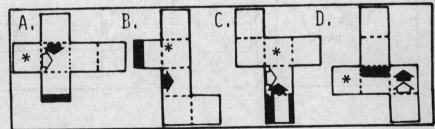

TEST V. BOX UNFOLDING

DIRECTIONS: Each question in this test consists of a numbered picture showing a box that is to be unfolded. If the box were unfolded it would look like one of the four cardboard patterns, lettered A, B, C, or D, in the other frames in the row. Choose the cardboard pattern that is unfolded from the lettered picture, and blacken your answer sheet accordingly.

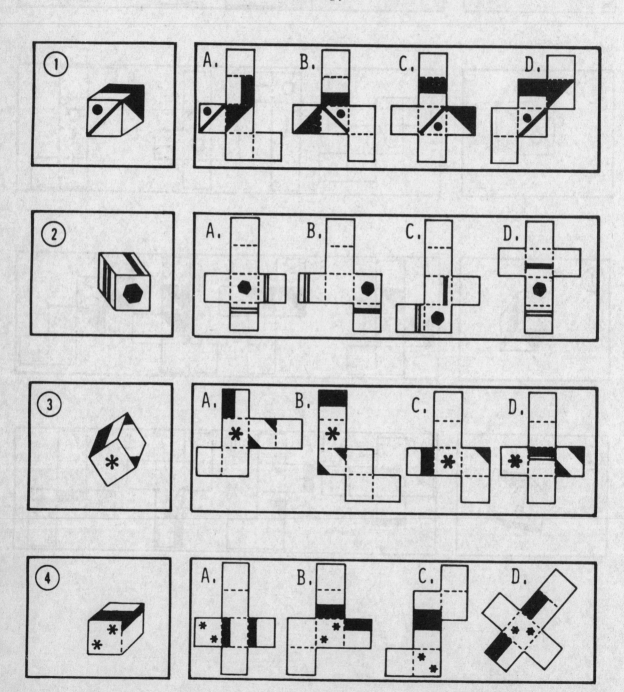

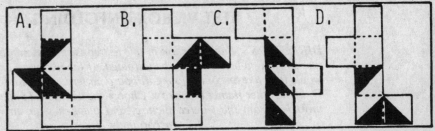

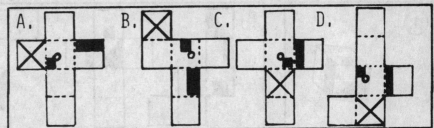

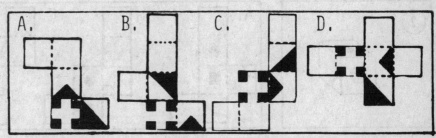

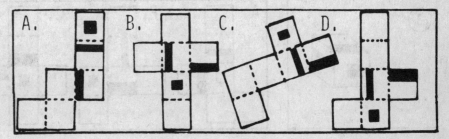

TEST VI. BOX UNFOLDING

DIRECTIONS: Each question in this test consists of a numbered picture showing a box that is to be unfolded. If the box were unfolded it would look like one of the four cardboard patterns, lettered A, B, C, or D, in the other frames in the row. Choose the cardboard pattern that is unfolded from the lettered picture, and blacken your answer sheet accordingly.

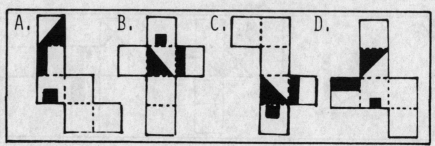

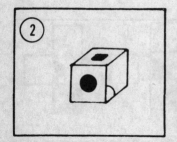
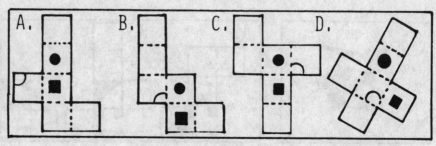

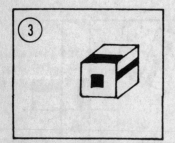
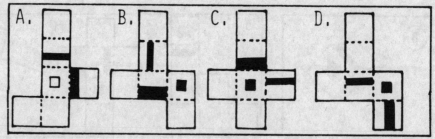

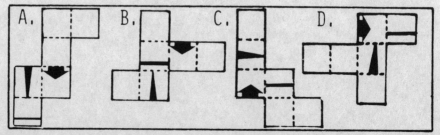

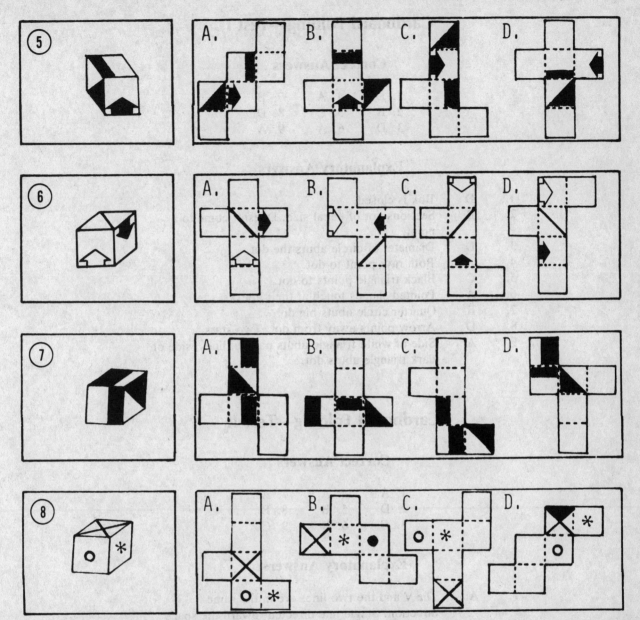

Cardboard Folding—Test I

Correct Answers

1. D	4. A	7. B
2. B	5. C	8. D
3. D	6. A	9. A

Explanatory Answers

1.	D	Box is closed.
2.	B	Sections not of equal size. Doesn't come to a point.
3.	D	Diameter of circle abuts the dot.
4.	A	Both rays point to dot.
5.	C	Black triangle points to dot.
6.	A	Pointed shapes touch at their apexes.
7.	B	Quarter circle abuts big dot.
8.	D	Arrow points away from dot. Two stars.
9.	A	Side of white triangle abuts parallel lines, side of dark triangle abuts dot.

Cardboard Folding—Test II

Correct Answers

1. A	4. A	7. A
2. D	5. B	8. B
3. C	6. D	9. C

Explanatory Answers

1.	A	The V and the two lines are in the same direction. Single line directed toward the space in the V.
2.	D	Dark half-face runs perpendicular to dark line.
3.	C	Cylinder with top and bottom.
4.	A	Short, square based pyramid.
5.	B	White half-face adjacent to diagonal stripe. Black square is on the white.
6.	D	Arrow points toward fat cross.
7.	A	Truncated cube.
8.	B	Rectangular solid with even-sided groove.
9.	C	Five sided base forms closed figure.

Cardboard Folding—Test III

Correct Answers

1. C	4. A	7. C
2. B	5. D	8. B
3. C	6. A	9. A

Explanatory Answers

1.	C	Triangular faces.
2.	B	Square base.
3.	C	Rectangular solid.
4.	A	Rectangular base, trapezoidal faces.
5.	D	Closed triangular prism.
6.	A	Closed. Pentagonal faces.
7.	C	Closed. Hexagonal faces.
8.	B	Open. Pentagonal faces, missing two faces.
9.	A	Rectangular solid, missing two faces.

Box Unfolding—Test IV

Correct Answers

1. B	5. C	
2. A	6. A	
3. D	7. B	
4. B	8. A	

Explanatory Answers

1.	B	Arrow points toward the hexagon.
2.	A	White half-face and solid side of black triangle abut the **I**.
3.	D	White half-face and black half-face abut the star.
4.	B	Black half-face abuts the star.
5.	C	Black-white half faces abut **I**.
6.	A	Neither ray points toward circle.
7.	B	Both arrows point toward black triangle's side.
8.	A	Black arrow points toward white half-face.

Box Unfolding—Test V

Correct Answers

1. D		5. D	
2. C		6. D	
3. A		7. D	
4. B		8. B	

Explanatory Answers

1.	D	White half-face abuts fully darkened circle.
2.	C	Dark stripe is perpendicular to hollow stripe.
3.	A	Black-white half-face abuts star perpendicularly.
4.	B	Black half-faces are perpendicular. Star in corner where they meet.
5.	D	Three black triangles, two share a side.
6.	D	The X does not abut the little square.
7.	D	Black triangle ▼ has vertex angle pointing toward cross.
8.	B	One black stripe has a large and small white area (not centered), the larger white space associated with this stripe abuts the face with the perpendicular black stripe.

Box Unfolding—Test VI

Correct Answers

1. B		5. D	
2. A		6. A	
3. C		7. C	
4. B		8. A	

Explanatory Answers

1.	B	Little black square abuts side of white triangle.
2.	A	Quarter open circle does *not* abut black square.
3.	C	Dark stripe directed toward black square. Other face is half black.
4.	B	Dark arrow points to side of triangle. Apex of triangle points toward center of black stripe. ("C" does not form a closed cube when folded.)
5.	D	Dark arrow points toward the face with the black-white split.
6.	A	White arrow points toward diagonal line, black points away from diagonal line.
7.	C	Dark sections do not share any common sides.
8.	A	All three designs abut one another.

FIGURE TURNING

There are two types of figure turning questions. In one type you are given a solid form and asked to pick the one of four alternatives which is the same form in a different position. In the other type you are given a cube which has a different design on each face and asked to choose the alternative or alternatives that could be the original cube after turning one or more times.

①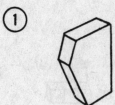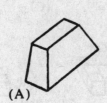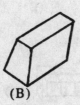

(A)   (B)   (C)   (D)

Sample question 1 illustrates a Solid Figure Turning question. Which of the alternatives lettered A, B, C and D represents figure 1 in a different position?

Figure 1 consists of a solid figure with 7 faces. Alternative D, which tilts the figure backward to expose the bottom surface, is the only alternative which could possibly represent the given figure in a different position.

Sample question 2 presents a Cube Turning problem. Look carefully at the numbered cube. Although you can see only 3 different designs, the cube actually has 6 different designs, one on each of its 6 faces. Now examine the 4 cubes lettered A, B, C and D. Select one or more of the lettered cubes according to the following rules:

1. If more than one of the lettered cubes could possibly be the cube on the left after turning, select the cube (or cubes) which is the cube on the left after one turn only.

2. If only one of the lettered cubes could be the cube on the left after turning, then that cube is the answer, regardless of how many turns have been made.

3. If more than one of the lettered cubes could possibly be the cube on the left after turning, but none of the lettered cubes could be the one on the left after one turn only, then select all possibilities as your answer.

②

(A)   (B)   (C)   (D)

Sample question 2 conforms to rule 2. Alternative A is wrong because the triangle on top should point toward the cross, not away from it. C is wrong because the cube would have to be turned upside down to put the triangle on the

right side of the cross, in which case the square would disappear. D is also wrong because the triangle should point toward the cross and in D it points toward a circle. B is the correct answer. It is the original cube turned twice.

81

TEST I. SOLID FIGURE TURNING

DIRECTIONS: Each numbered figure is made up of cubes or other forms which are assumed to be glued together. Next to each numbered figure are four lettered figures. Choose the one lettered figure (A, B, C or D) which is the numbered figure turned in a different position. In order to select the correct answer, you may have to mentally turn figures over, turn them around or turn them both over and around.

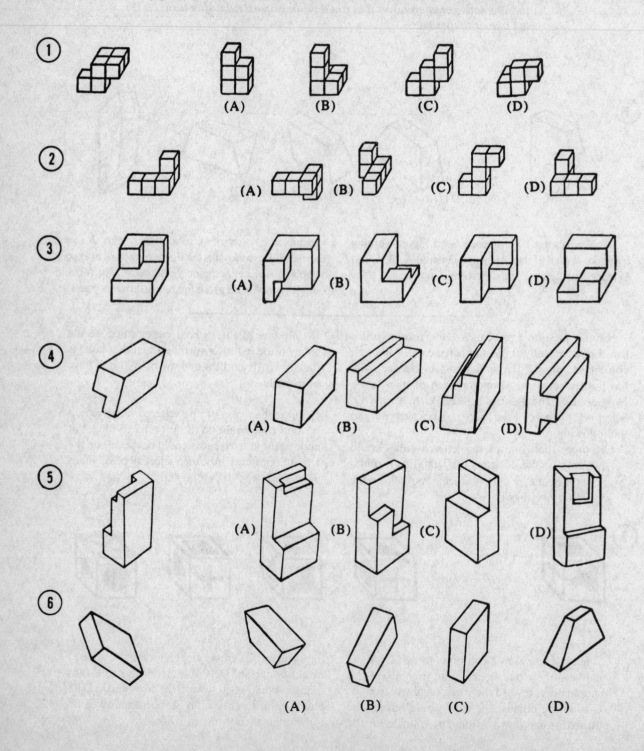

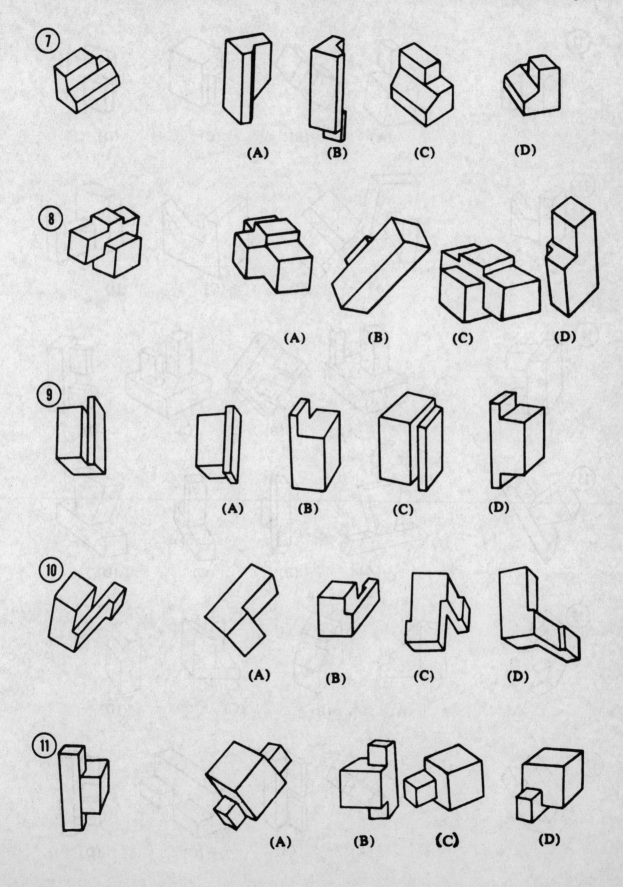

7 (A) (B) (C) (D)

8 (A) (B) (C) (D)

9 (A) (B) (C) (D)

10 (A) (B) (C) (D)

11 (A) (B) (C) (D)

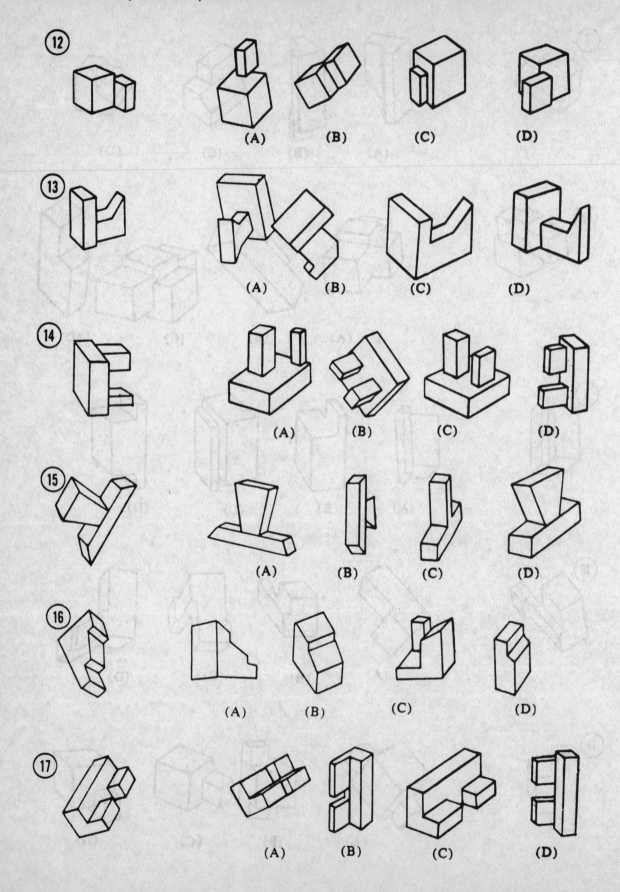

12

(A) (B) (C) (D)

13

(A) (B) (C) (D)

14

(A) (B) (C) (D)

15

(A) (B) (C) (D)

16

(A) (B) (C) (D)

17

(A) (B) (C) (D)

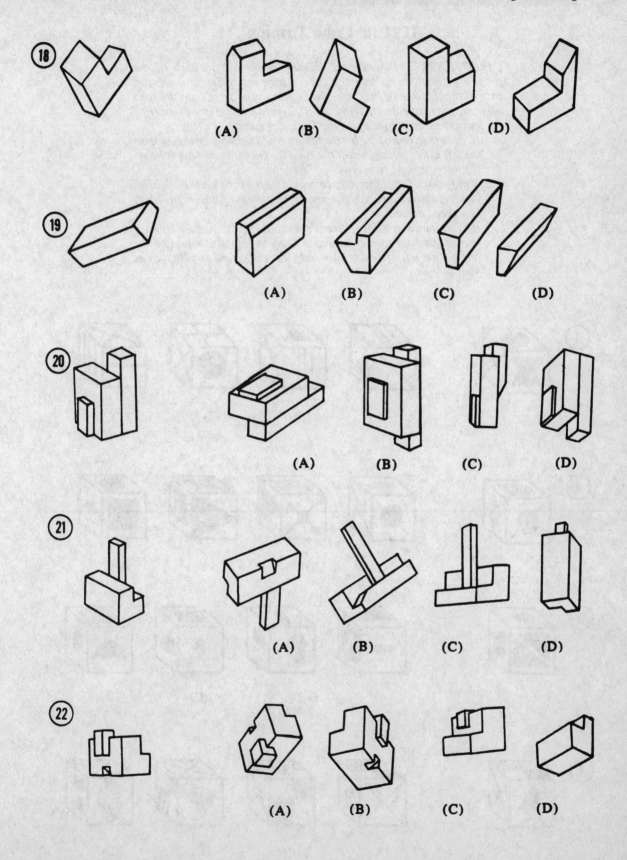

TEST II. Cube Turning

DIRECTIONS: The numbered drawing at the left in each of the following questions represents a cube. There is a different design on each of the six faces of the cube. Next to each numbered cube are four other drawings of cubes lettered, A, B, C and D. Select one or more of the lettered cubes according to the following rules:

1. *If more than one of the lettered cubes could possibly be the cube on the left after turning, select the cube (or cubes) that is the cube on the left after one turn only.*

2. *If only one of the lettered cubes could be the cube on the left after turning, then that cube is the answer, regardless of how many turns have been made.*

3. *If more than one of the lettered cubes could possibly be the cube on the left after turning, but none of the lettered cubes could be the one on the left after one turn only, then select all possibilities as your answer.*

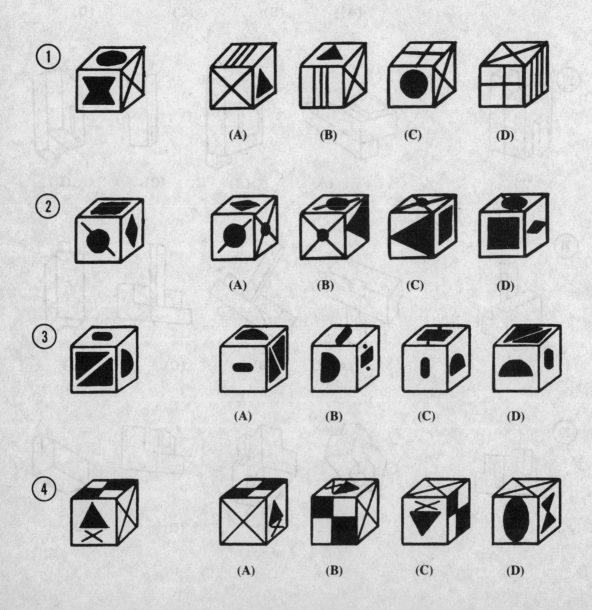

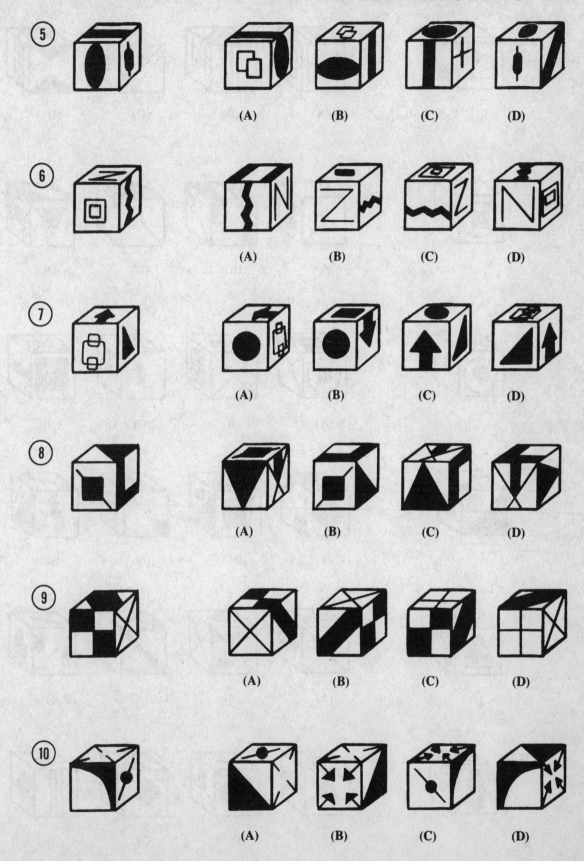

(5)
(A) (B) (C) (D)

(6)
(A) (B) (C) (D)

(7)
(A) (B) (C) (D)

(8)
(A) (B) (C) (D)

(9)
(A) (B) (C) (D)

(10)
(A) (B) (C) (D)

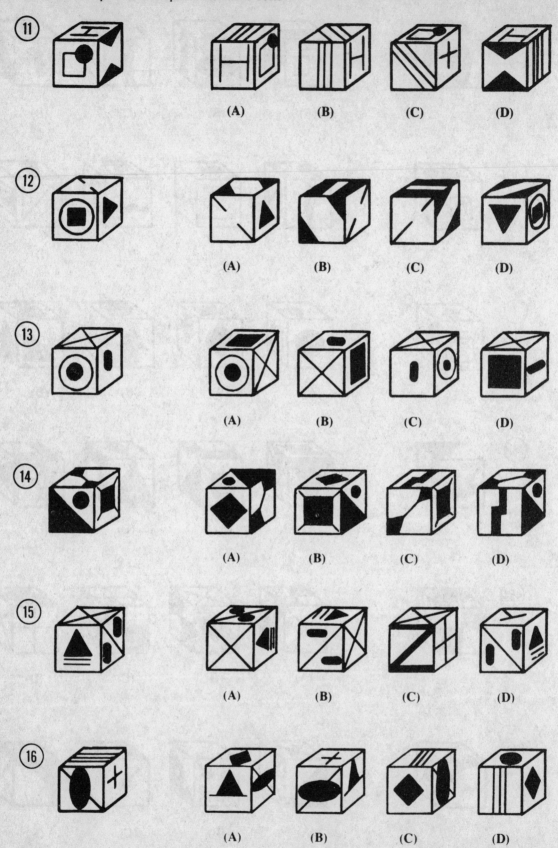

11   (A)   (B)   (C)   (D)

12   (A)   (B)   (C)   (D)

13   (A)   (B)   (C)   (D)

14   (A)   (B)   (C)   (D)

15   (A)   (B)   (C)   (D)

16   (A)   (B)   (C)   (D)

⑰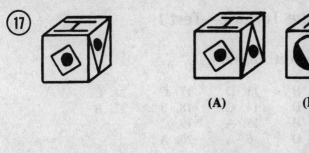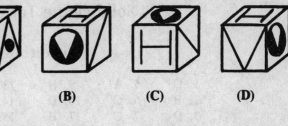

(A)   (B)   (C)   (D)

⑱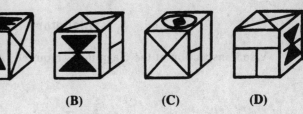

(A)   (B)   (C)   (D)

Solid Figure Turning—Test I

Correct Answers

1. D	5. A	9. B	13. D	17. C	21. C
2. B	6. C	10. D	14. C	18. B	22. B
3. D	7. C	11. B	15. A	19. C	
4. B	8. C	12. D	16. A	20. A	

Explanatory Answers

Visual inspection of the diagrams is required. Answers are self- explanatory.

Cube Turning—Test II

Correct Answers

1. C	4. D	7. A,C	10. B	13. A	16. B,C
2. A	5. B	8. A,D	11. D	14. D	17. C
3. B	6. B	9. C	12. A	15. B,C	18. A

Explanatory Answers

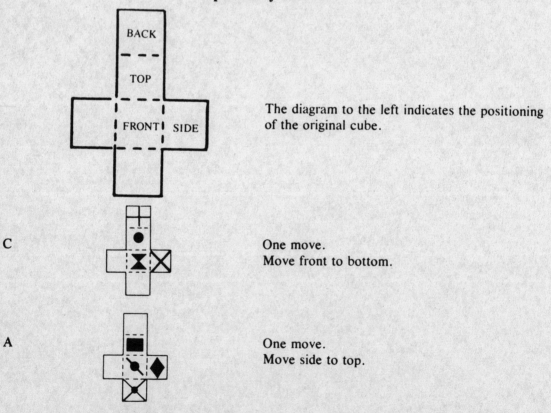

The diagram to the left indicates the positioning of the original cube.

1. C

One move.
Move front to bottom.

2. A

One move.
Move side to top.

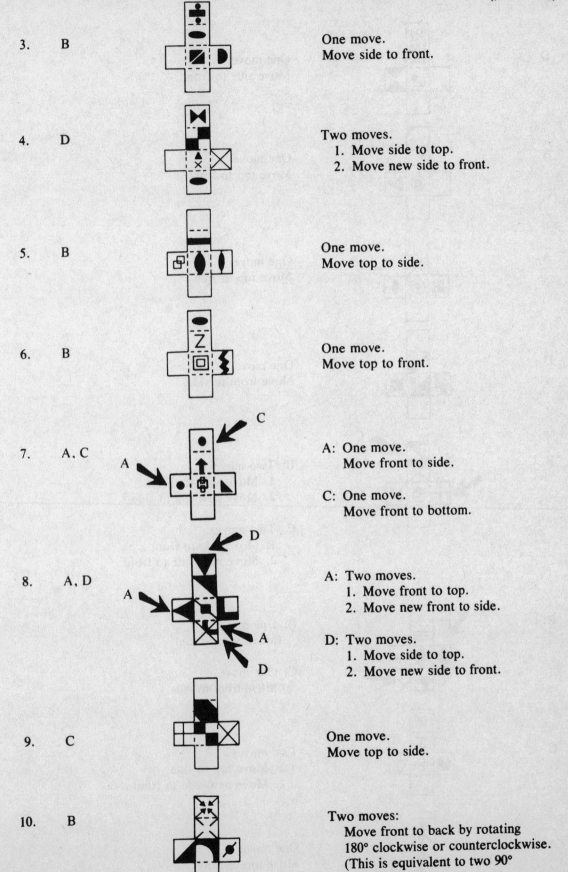

3. B

One move.
Move side to front.

4. D

Two moves.
1. Move side to top.
2. Move new side to front.

5. B

One move.
Move top to side.

6. B

One move.
Move top to front.

7. A, C

A: One move.
Move front to side.

C: One move.
Move front to bottom.

8. A, D

A: Two moves.
1. Move front to top.
2. Move new front to side.

D: Two moves.
1. Move side to top.
2. Move new side to front.

9. C

One move.
Move top to side.

10. B

Two moves:
Move front to back by rotating
180° clockwise or counterclockwise.
(This is equivalent to two 90°
rotations.)

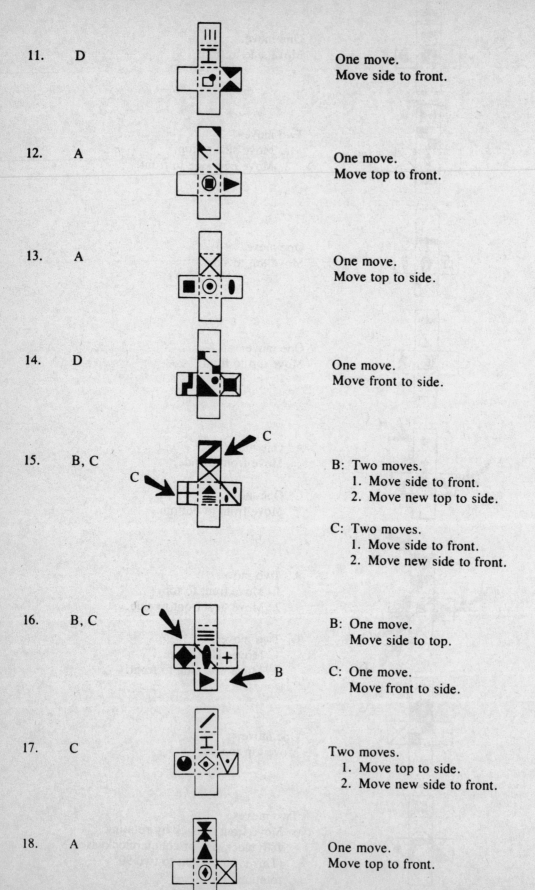

11. D

One move.
Move side to front.

12. A

One move.
Move top to front.

13. A

One move.
Move top to side.

14. D

One move.
Move front to side.

15. B, C

B: Two moves.
   1. Move side to front.
   2. Move new top to side.

C: Two moves.
   1. Move side to front.
   2. Move new side to front.

16. B, C

B: One move.
   Move side to top.

C: One move.
   Move front to side.

17. C

Two moves.
   1. Move top to side.
   2. Move new side to front.

18. A

One move.
Move top to front.

PERCEPTUAL ABILITY TESTS

Perceptual Ability Tests provide a measure of your powers of observation by requiring you to compare, contrast and rank similar figures. These abilities are best tested by the following questions types: Similarities and Differences Between Objects, Line Ranking, Angle Comparisons. This chapter provides detailed explanations of each question type followed by practice tests to help you improve your perceptual skill.

SIMILARITIES AND DIFFERENCES BETWEEN OBJECTS

This is a test of your powers of observation. Each question consists of five drawings lettered A, B, C, D and E. Four of the drawings are exactly alike; one is slightly different. The task is to find the one drawing that is different from the other four. Mark your answer sheet for the letter of the drawing that is different.

A SAMPLE QUESTION EXPLAINED

At first glance, the five drawings below may seem identical. However, careful observation will reveal that one of the drawings is slightly different from the other four. Can you spot the drawing that is different?

The drawing lettered C is the one that is different. There is a line missing from the right corner segment of this drawing which is present in each of the other four drawings in the group.

| (A) | (B) | (C) | (D) | (E) |

LINE RANKING AND ANGLE COMPARISON TESTS

The Line Ranking and Angle Comparison Tests are designed to measure your two-dimensional perception and problem-solving ability. These are question styles which are probably unfamiliar and with which you have had no previous experience. While a naturally 'good eye' will stand you in good stead, you can prepare yourself for these questions by learning how to approach each problem and by investing sufficient time in practice.

A LINE RANKING QUESTION EXPLAINED

Choose the alternative which correctly ranks the lines from the shortest to the longest.

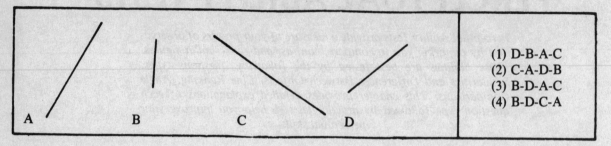

(1) D-B-A-C
(2) C-A-D-B
(3) B-D-A-C
(4) B-D-C-A

Answer: 3

Hints for Ranking Lines:

1. First look at the answers. By simple inspection you may be able to eliminate one or two choices. Often the shortest and longest lines are very obvious.

2. Once you have narrowed your choices, note which alternatives are offered and concentrate on eliminating those which are not correct. You want to avoid wasting time working out your own ranking, only to find that your order is not offered.

3. Compare adjacent lines to work out relative rankings; distance distorts.

4. Feel free to move your paper around. Turn the test booklet to get a different perspective on the lines which puzzle you.

5. When comparing lines with reference to a specific point on the page, be certain that the point of reference really applies to both lines being compared (for example, that both lines really begin at the same distance from the box line).

6. Be aware of optical illusions. Diagonal lines appear to be shorter than vertical or horizontal lines of the same length.

AN ANGLE COMPARISON QUESTION EXPLAINED

For each question, find the one angle (numbered 1 through 8) which is the same
as angle 'x' in the figure on the left.

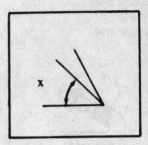 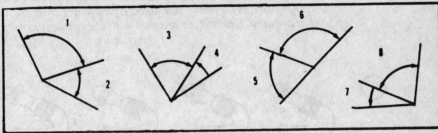

Answer: 2

Hints for Comparing Angles:

1. Look at one angle at a time; the juxtaposition of angles can only confuse you. Cover one angle with your hand or edge of your answer sheet while concentrating on its partner.

2. Concern yourself only with the angle. Disregard the lengths of the arms; they are irrelevant and distracting.

3. Try to focus on the angle itself, ignoring the numbered arc. You may find it helpful to draw an imaginary arc at the same location on each angle that you are comparing.

4. Turn your booklet at will to view all angles in the same position. You will be amazed at the insight you gain when you see each angle from the same perspective as the original angle.

As in taking all tests: Do not spend too much time on any one question. If you are stumped by one question go on to the next one and come back only if you have time. All the questions count equally, so it is to your advantage to answer as many of them as possible.

TEST I. SIMILARITIES AND DIFFERENCES BETWEEN OBJECTS

DIRECTIONS: Each question consists of five lettered drawings. Four of the drawings are exactly alike; one is slightly different. Select the one drawing that is different and mark your answer sheet for the letter of that drawing. Check your answers with the correct answers at the end of the chapter.

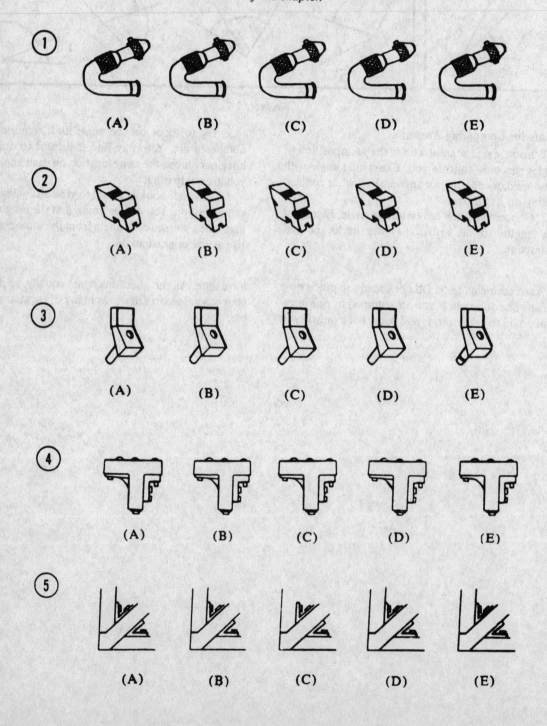

1

(A)   (B)   (C)   (D)   (E)

2

(A)   (B)   (C)   (D)   (E)

3

(A)   (B)   (C)   (D)   (E)

4

(A)   (B)   (C)   (D)   (E)

5

(A)   (B)   (C)   (D)   (E)

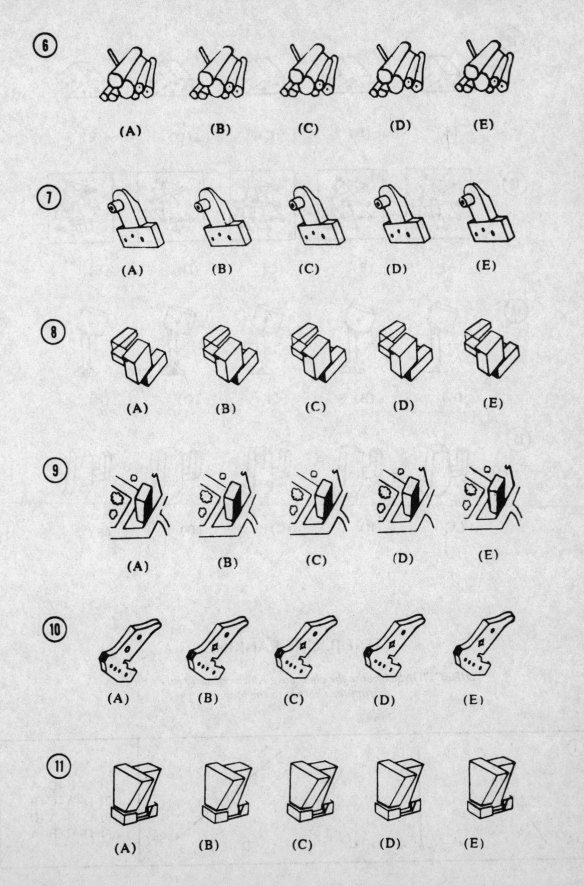

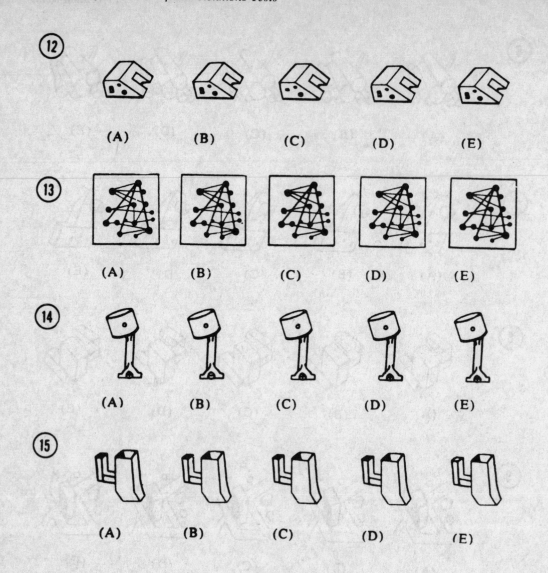

⑫

(A)   (B)   (C)   (D)   (E)

⑬

(A)   (B)   (C)   (D)   (E)

⑭

(A)   (B)   (C)   (D)   (E)

⑮

(A)   (B)   (C)   (D)   (E)

TEST II. LINE RANKING

DIRECTIONS: Choose the alternative which correctly ranks the lines from the shortest to the longest.

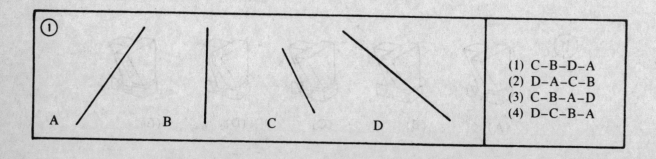

①   A   B   C   D

(1) C–B–D–A
(2) D–A–C–B
(3) C–B–A–D
(4) D–C–B–A

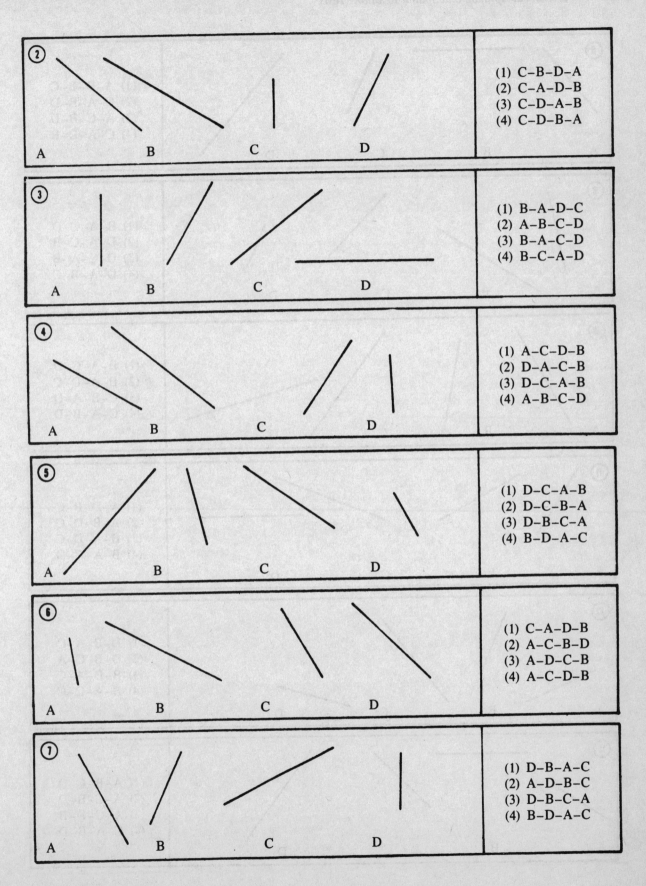

2

(1) C–B–D–A
(2) C–A–D–B
(3) C–D–A–B
(4) C–D–B–A

A   B   C   D

3

(1) B–A–D–C
(2) A–B–C–D
(3) B–A–C–D
(4) B–C–A–D

A   B   C   D

4

(1) A–C–D–B
(2) D–A–C–B
(3) D–C–A–B
(4) A–B–C–D

A   B   C   D

5

(1) D–C–A–B
(2) D–C–B–A
(3) D–B–C–A
(4) B–D–A–C

A   B   C   D

6

(1) C–A–D–B
(2) A–C–B–D
(3) A–D–C–B
(4) A–C–D–B

A   B   C   D

7

(1) D–B–A–C
(2) A–D–B–C
(3) D–B–C–A
(4) B–D–A–C

A   B   C   D

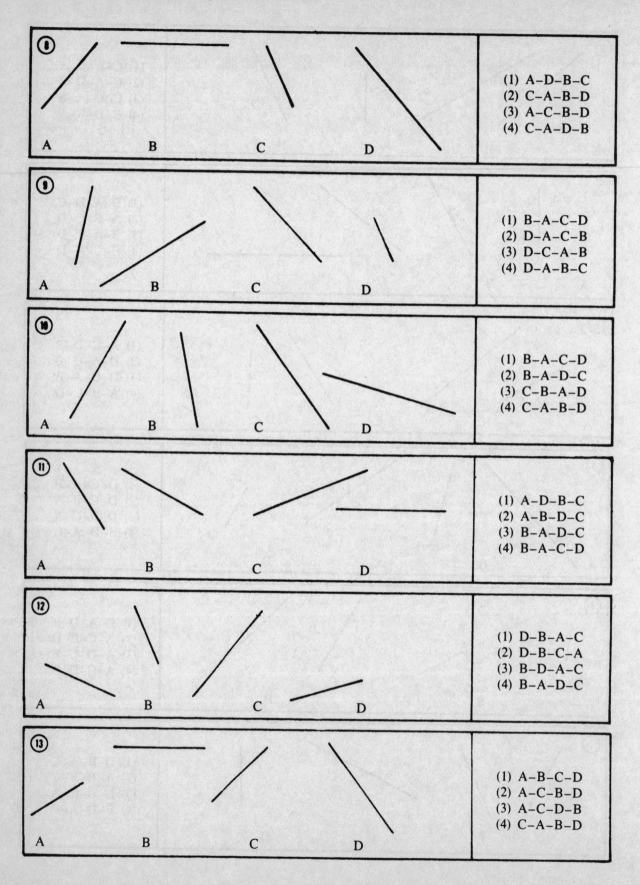

8

(1) A–D–B–C
(2) C–A–B–D
(3) A–C–B–D
(4) C–A–D–B

9

(1) B–A–C–D
(2) D–A–C–B
(3) D–C–A–B
(4) D–A–B–C

10

(1) B–A–C–D
(2) B–A–D–C
(3) C–B–A–D
(4) C–A–B–D

11

(1) A–D–B–C
(2) A–B–D–C
(3) B–A–D–C
(4) B–A–C–D

12

(1) D–B–A–C
(2) D–B–C–A
(3) B–D–A–C
(4) B–A–D–C

13

(1) A–B–C–D
(2) A–C–B–D
(3) A–C–D–B
(4) C–A–B–D

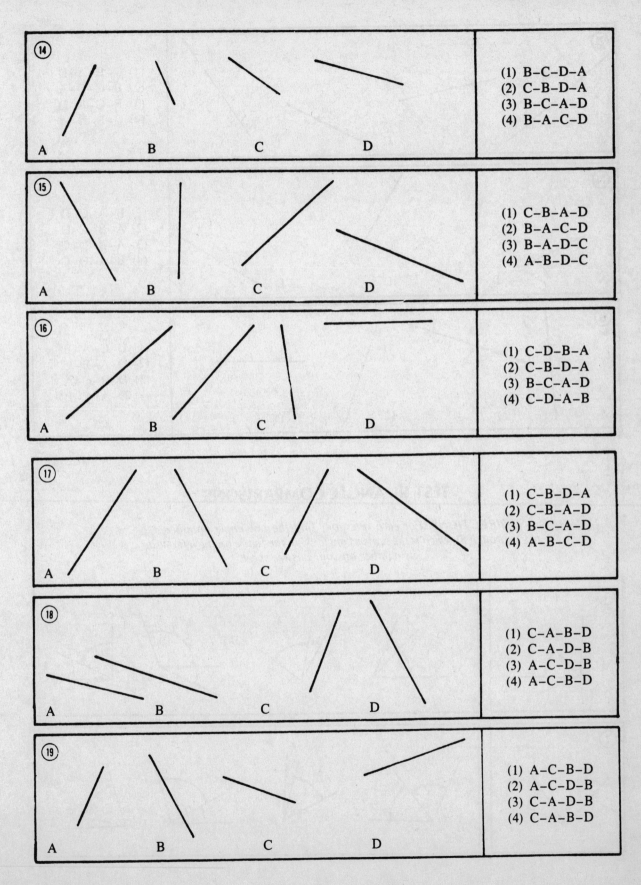

14

(1) B–C–D–A
(2) C–B–D–A
(3) B–C–A–D
(4) B–A–C–D

15

(1) C–B–A–D
(2) B–A–C–D
(3) B–A–D–C
(4) A–B–D–C

16

(1) C–D–B–A
(2) C–B–D–A
(3) B–C–A–D
(4) C–D–A–B

17

(1) C–B–D–A
(2) C–B–A–D
(3) B–C–A–D
(4) A–B–C–D

18

(1) C–A–B–D
(2) C–A–D–B
(3) A–C–D–B
(4) A–C–B–D

19

(1) A–C–B–D
(2) A–C–D–B
(3) C–A–D–B
(4) C–A–B–D

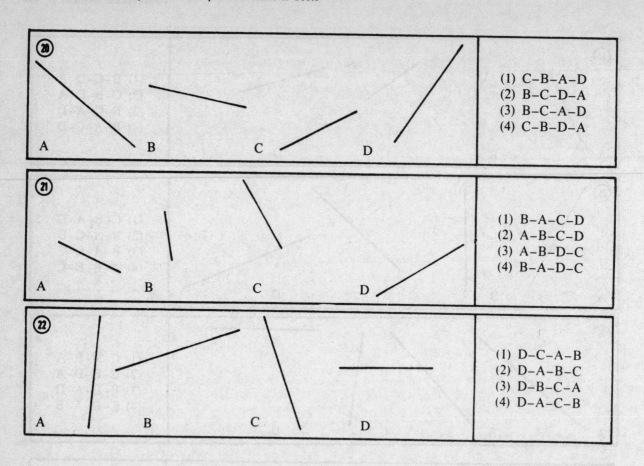

20

(1) C–B–A–D
(2) B–C–D–A
(3) B–C–A–D
(4) C–B–D–A

21

(1) B–A–C–D
(2) A–B–C–D
(3) A–B–D–C
(4) B–A–D–C

22

(1) D–C–A–B
(2) D–A–B–C
(3) D–B–C–A
(4) D–A–C–B

TEST III. ANGLE COMPARISONS

DIRECTIONS: For each question, find the one angle (numbered 1 through 8) which is the same as angle 'x' in the figure on the left. Write its number on your answer sheet.

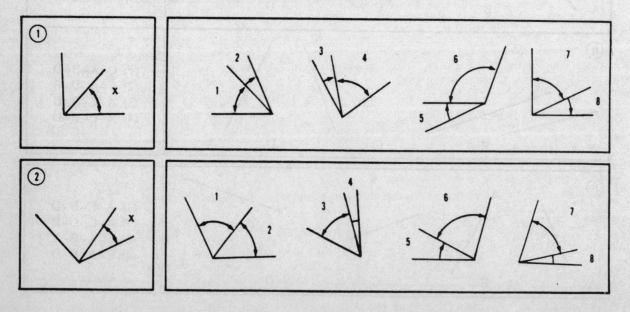

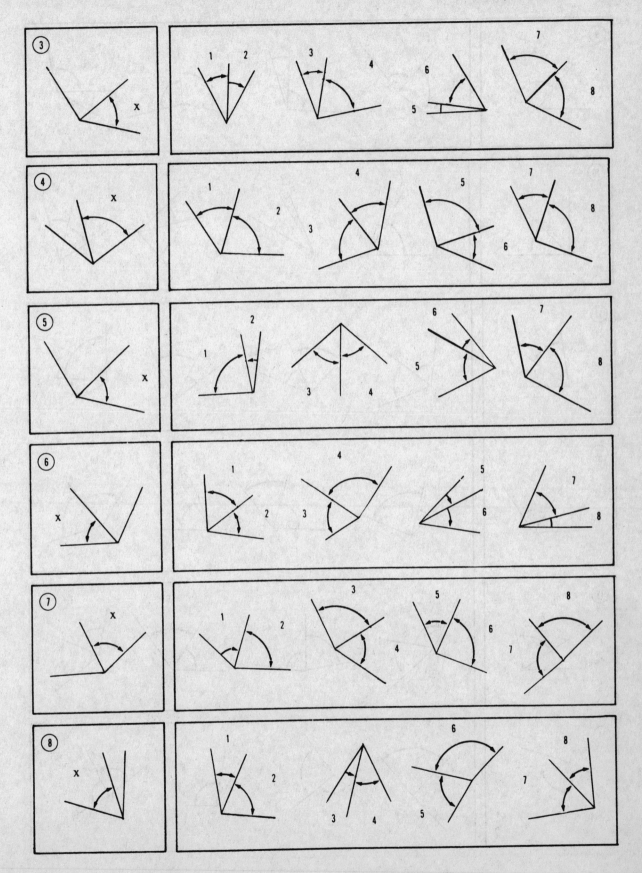

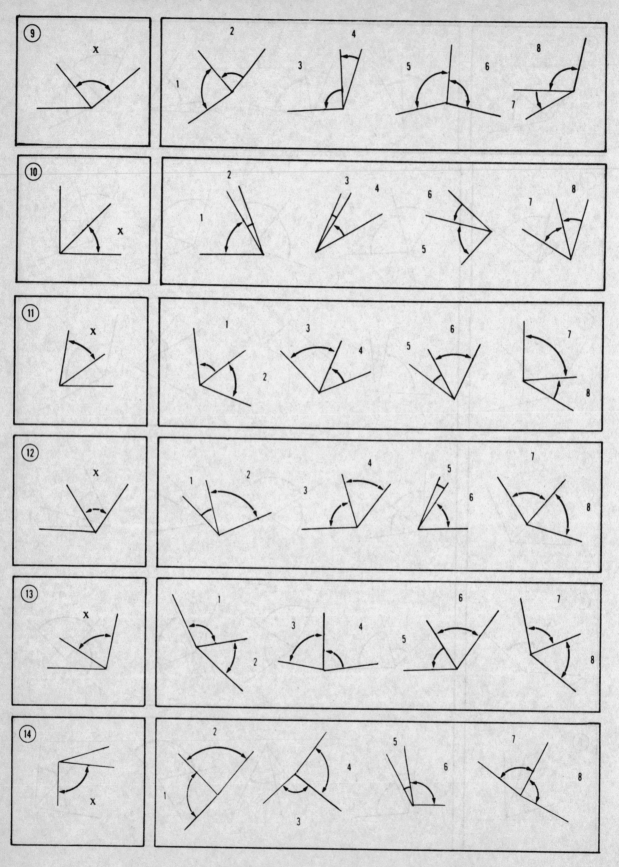

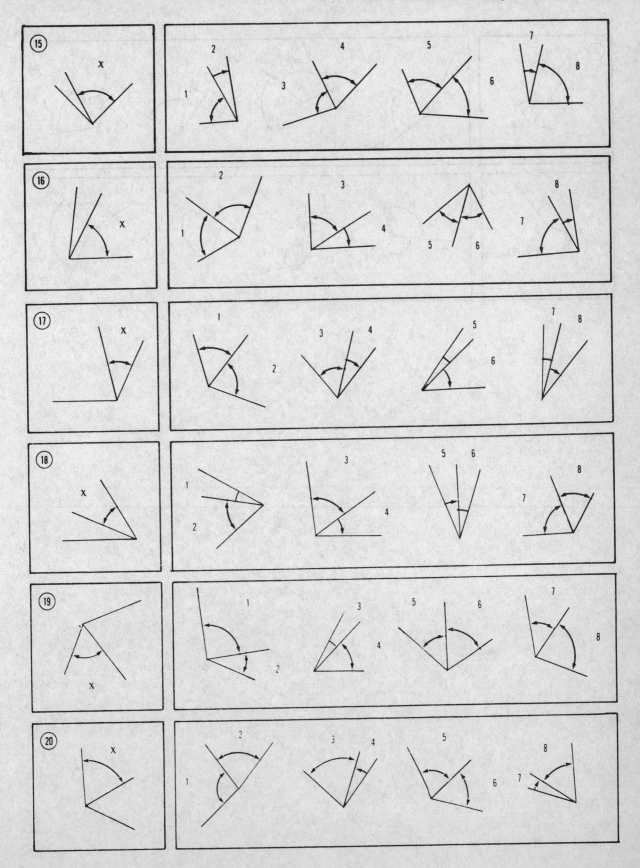

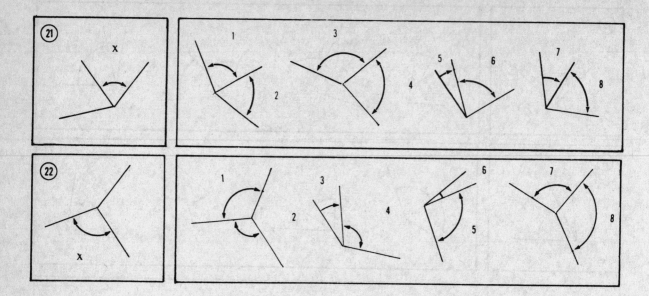

Similarities and Differences Between Objects—Test I

Correct Answers

1. D	4. E	7. B	10. A	13. B
2. B	5. C	8. A	11. B	14. B
3. E	6. D	9. D	12. C	15. E

Explanatory Answers

Arrow points to the position of the particular difference.

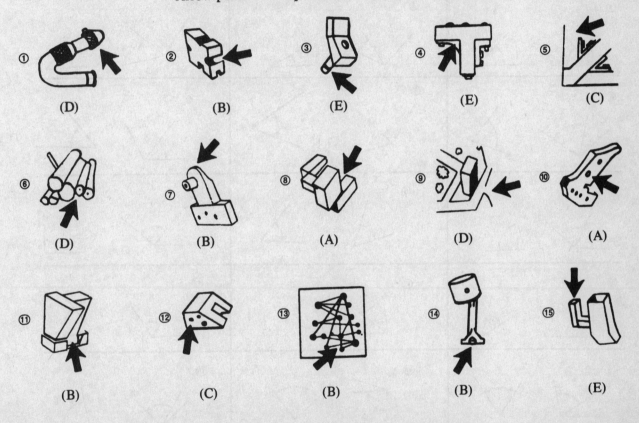

Line Ranking—Test II

Correct Answers

1. 3	5. 3	9. 2	13. 2	17. 2	21. 1
2. 3	6. 4	10. 1	14. 3	18. 2	22. 4
3. 2	7. 1	11. 2	15. 2	19. 1	
4. 3	8. 2	12. 3	16. 1	20. 4	

Explanatory Answers

A ruler can be used to discern the difference between the lengths of the line segments.

Angle Comparisons—Test III

Correct Answers

1. 1	5. 5	9. 3	13. 6	17. 6	21. 6
2. 5	6. 7	10. 7	14. 2	18. 4	22. 4
3. 6	7. 2	11. 4	15. 8	19. 6	
4. 3	8. 7	12. 8	16. 3	20. 3	

Explanatory Answers

VISUAL-MOTOR COORDINATION TESTS

Visual-Motor Coordination Tests are tests of speed and accuracy. They are meant to explore your facility in dealing with things quickly and correctly. Among the more usual tests of Visual-Motor Coordination are Letter-Symbol Coding, Inspection Tests, Examining Objects and Mazes and Pursuits. The following pages present samples of each of these question types. Answers to all of these tests are at the end of the chapter.

TEST I. LETTER-SYMBOL CODING

DIRECTIONS: This test consists of 100 items, all of which are based on the Letter-Symbol Code shown below. The Letter-Symbol Code consists of 10 divided rectangles. In the upper half of each rectangle there is a letter. In the lower half there is a symbol. This Code is followed by 100 rectangles divided into 3 parts. The upper third is the item number. The middle third contains a symbol (one of the ten symbols which form the Letter-Symbol Code). The bottom third, entitled Letter Answer, is where you will record your answers by inserting the letter that corresponds to the symbol shown. Work quickly but accurately. Allow yourself only 5 minutes to complete the entire test. The first 10 items are answered correctly to show you how to proceed with this type of test.

LETTER-SYMBOL CODE FOR ITEMS 1 TO 100

LETTER	A	B	C	D	E	F	G	H	I	J
SYMBOL	⌣	▽	⊓	⌒	⩑	⋈	⊢⊣	⨯	▽	▽

ITEMS 1 TO 100. THE FIRST TEN ARE SAMPLES, ANSWERED CORRECTLY.

ITEM NUMBER	1	2	3	4	5	6	7	8	9	10
SYMBOL	⋈	⨯	▽	▽	⊓	⊢⊣	⌒	⌣	⩑	▽
LETTER ANSWER	**F**	**H**	**I**	**J**	**C**	**G**	**D**	**A**	**E**	**B**
ITEM NUMBER	11	12	13	14	15	16	17	18	19	20
SYMBOL	⊢⊣	⩑	▽	⨯	▽	⋈	⊓	▽	▽	⌣
LETTER ANSWER										
ITEM NUMBER	21	22	23	24	25	26	27	28	29	30
SYMBOL	⌣	▽	⨯	⊢⊣	▽	⊓	⩑	⋈	⌒	▽
LETTER ANSWER										

ITEM NUMBER	31	32	33	34	35	36	37	38	39	40
SYMBOL	⊓	◹	M	⋈	▽	⊠	⊢○⊣	◡	▽	⟋
LETTER ANSWER										

ITEM NUMBER	41	42	43	44	45	46	47	48	49	50
SYMBOL	⟋	⊢○⊣	◡	▽	M	◹	⊠	⊓	⋈	▽
LETTER ANSWER										

ITEM NUMBER	51	52	53	54	55	56	57	58	59	60
SYMBOL	⋈	M	▽	◹	⊓	⊠	◡	⟋	⊢○⊣	▽
LETTER ANSWER										

ITEM NUMBER	61	62	63	64	65	66	67	68	69	70
SYMBOL	◡	⟋	⊠	M	⊢○⊣	▽	◹	▽	⊓	⋈
LETTER ANSWER										

ITEM NUMBER	71	72	73	74	75	76	77	78	79	80
SYMBOL	▽	⋈	◹	⊓	◡	▽	⊠	⊢○⊣	⟋	M
LETTER ANSWER										

ITEM NUMBER	81	82	83	84	85	86	87	88	89	90
SYMBOL	◹	⊓	⊢○⊣	▽	⟋	⋈	▽	◡	M	⊠
LETTER ANSWER										

ITEM NUMBER	91	92	93	94	95	96	97	98	99	100
SYMBOL	▽	⟋	⋈	◹	▽	◡	M	⊓	⊠	⊢○⊣
LETTER ANSWER										

TEST II. COUNTING CROSSES
AND ZEROS

DIRECTIONS: In this test each question asks for the number of X's or O's or both in a certain area of the diagram shown. Count the X's as directed. If an X or an O falls across a line dividing one area from another, count it as being in the area in which most of it falls. Look for your answer among those suggested and mark the answer column accordingly. If your answer is not listed, mark E.

	COLUMN A	COLUMN B	COLUMN C	COLUMN D	COLUMN E
ROW 1	X X X X X X O O X O O OO O O O X O X X X X O X O O O O O O X	O X X O X O X O O X O O O X X X X X X O O O X O O X X X O X	X O X O O X X X O X X O O X O O O X X O X O X O O X X X O X	O O O X X X X X O XX X X X X O O O O O O O O O O O O O O	X X X X O O O O X X X O X O O O O XX O X O X O OO X X X O O XO
ROW 2	O X O X O O X X X O X O O O O O O O X X X X X O O O X X X O	X O O O X O O X X X O X X X X X X X X O O O O O X O X O	X O X O X O X O X O X X X O O O O O X X X X X X O	X O X O O O X O O X X O O O X O X X X O X O X O O	X O O O O O X X X X X O O O X X O O X O X X O O O
ROW 3	X O O O X O O X X X O X X O O O X O O O O O O X X X X X X X	X X X X O O O O X X X X X O O O O O X X X X X O O	X O O X O O O X X O X O X X O O O X O O X X X O X X O X	X O O O O O X X X X O X X X X X O O O X O X O X O	X X X X X O O O O X X X X X X O O O O O X O O O X O
ROW 4	O O O O O O X X X O X O O X O X O X X O O X O O O O X X O	O O X X X X X O X X O O O O X O O X O X X X X X X	X X X O O X O O O X X X X O O X O O O X X O X X O O O	X O O X X X O O O O O X O X O X X O X O X O X X O	O X X O X O X X X O X O O O O O X X X X X O O X X X O

1. How many X's are there in Row 1, Column B?
   (A) 14
   (B) 16
   (C) 17
   (D) 18
   (E) None of these

2. How many O's in Row 1, Columns A and B?
   (A) 30
   (B) 31
   (C) 33
   (D) 34
   (E) None of these

3. How many X's in Row 2, Column A?
   (A) 10
   (B) 11
   (C) 12
   (D) 14
   (E) None of these

4. How many O's in Row 3, Column C?
   (A) 12
   (B) 13
   (C) 15
   (D) 16
   (E) None of these

5. How many X's and O's in Row 3, Column B?
   (A) 25
   (B) 26
   (C) 27
   (D) 29
   (E) None of these

6. How many X's in Row 4, Column A?
   (A) 13
   (B) 14
   (C) 16
   (D) 17
   (E) None of these

7. How many X's in Row 2, Column D?
   (A) 11
   (B) 12
   (C) 14
   (D) 15
   (E) None of these

8. How many O's in Row 2, Column C?
   (A) 10
   (B) 11
   (C) 12
   (D) 14
   (E) None of these

9. How many X's and O's in Row 3, Column E?
   (A) 24
   (B) 25
   (C) 26
   (D) 22
   (E) None of these

10. How many O's in Row 3, Column E?
    (A) 10
    (B) 11
    (C) 12
    (D) 13
    (E) None of these

11. How many X's in Row 1, Column E?
    (A) 16
    (B) 17
    (C) 19
    (D) 20
    (E) None of these

12. How many O's in Row 4, Col. D?
    (A) 11
    (B) 12
    (C) 14
    (D) 15
    (E) None of these

13. How many O's in Rows 2 and 3, Column B?
    (A) 22
    (B) 23
    (C) 24
    (D) 25
    (E) None of these

14. How many X's in Row 4, Columns D and E?
    (A) 25
    (B) 26
    (C) 27
    (D) 28
    (E) None of these

15. How many X's and O's in Row 2, Column C?
    (A) 25
    (B) 26
    (C) 27

(D) 28
(E) None of these

16. How many O's in Row 2, Column E?
    (A) 13
    (B) 14
    (C) 15
    (D) 16
    (E) None of these

17. How many O's in Row 4, Column A?
    (A) 17
    (B) 18
    (C) 20
    (D) 21
    (E) None of these

18 How many X's in Rows 2 and 3, Column C?
    (A) 27
    (B) 28
    (C) 29
    (D) 31
    (E) None of these

19. How many O's in Row 4, Columns B and C?
    (A) 21
    (B) 22
    (C) 23
    (D) 24
    (E) None of these

20. How many X's and O's in Row 3, Columns D and E?
    (A) 49
    (B) 50
    (C) 51
    (D) 52
    (E) None of these

21. How many O's in Row 1, Column D?
    (A) 13
    (B) 14
    (C) 15
    (D) 16
    (E) None of these

22. How many X's in Row 4, Column B?
    (A) 15
    (B) 16
    (C) 18
    (D) 19
    (E) None of these

23. How many O's in Rows 2, 3 and 4, Column B?
    (A) 30
    (B) 31
    (C) 32
    (D) 33
    (E) None of these

24. How many X's in Row 3, Columns C, D, and E?
(A) 39
(B) 40
(C) 42
(D) 43
(E) None of these

TEST III. EXAMINING OBJECTS

DIRECTIONS: The diagram below shows five bins into which pairs of parts are to be sorted.

Bin A is for pairs that are ALL RIGHT. One part is a square, the other round, and both have the same number.

Bin B is for pairs that have a part that is BASHED IN. One part is square, the other round, and both have the same number.

Bin C is for a COMBINATION of two round or two square parts, both bearing the same number.

Bin D is for pairs in which the parts have DIFFERENT NUMBERS. One part is round and the other square.

Bin E is for pairs in which the number spaces are EMPTY. One part is round and the other part square.

For each question, examine the pair of parts shown and decide which bin they go into. Record the letter of that bin on your answer sheet. The first 3 practice problems are answered correctly to show you how this test is done.

Code for Inspection Test

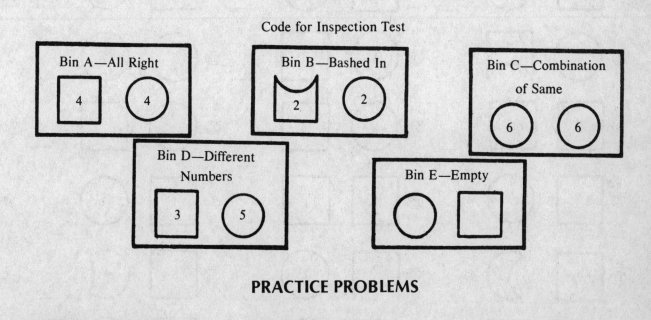

PRACTICE PROBLEMS

① 2 2 ... ___ ⑬ 6 6 ... ___ ㉕ 8 8 ... ___

② 3 3 ... ___ ⑭ 3 3 ... ___ ㉖ 2 2 ... ___

③ 5 5 ... ___ ⑮ ___ ... ___ ㉗ ___ ... ___

④ 3 7 ... ___ ⑯ 5 5 ... ___ ㉘ 4 4 ... ___

⑤ 1 1 ... ___ ⑰ 7 7 ... ___ ㉙ 5 3 ... ___

⑥ 4 4 ... ___ ⑱ 4 4 ... ___ ㉚ 7 7 ... ___

⑦ 1 1 ... ___ ⑲ ___ ... ___ ㉛ 1 1 ... ___

⑧ 7 7 ... ___ ⑳ 1 3 ... ___ ㉜ 2 2 ... ___

⑨ 6 6 ... ___ ㉑ 8 8 ... ___ ㉝ ___ ... ___

⑩ ___ ... ___ ㉒ 1 1 ... ___ ㉞ 1 3 ... ___

⑪ 3 3 ... ___ ㉓ 2 2 ... ___ ㉟ 6 6 ... ___

⑫ 8 5 ... ___ ㉔ 5 7 ... ___ ㊱ 3 3 ... ___

TEST IV. MAZES AND PURSUITS

DIRECTIONS: In this test, your speed and accuracy are tested by requiring you to follow a path quickly and accurately to its correct end. At the left of each pursuit diagram you will find a series of numbers, each corresponding to a line. Trace each numbered line to its correct end. Each one ends at a letter. Mark your answer sheet for the letter at which the numbered line ends. For example, the answer to question 1 is D as illustrated by the dark line connecting the number 1 with the letter D.

PURSUITS

MAZES

DIRECTIONS: Follow each numbered path to its end. Mark your answer sheet for the letter of the exit which allows you to pass through the maze without crossing any solid line.

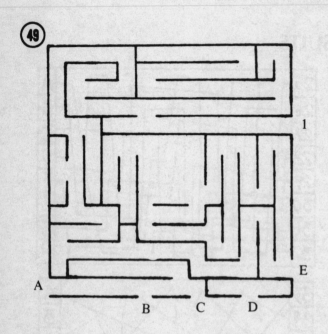

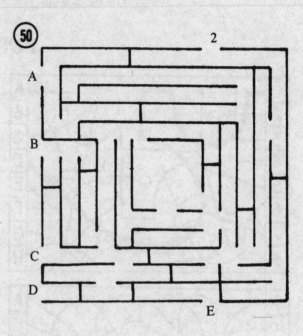

Letter-Symbol Coding—Test I

Correct Answers

1. F	26. C	51. E	76. J
2. H	27. E	52. F	77. H
3. I	28. F	53. J	78. G
4. J	29. D	54. I	79. D
5. C	30. I	55. C	80. F
6. G	31. C	56. H	81. I
7. D	32. I	57. A	82. C
8. A	33. F	58. D	83. G
9. E	34. E	59. G	84. J
10. B	35. J	60. B	85. D
11. G	36. H	61. A	86. E
12. E	37. G	62. D	87. B
13. D	38. A	63. H	88. A
14. H	39. B	64. F	89. F
15. I	40. D	65. G	90. H
16. F	41. D	66. B	91. J
17. C	42. G	67. I	92. D
18. J	43. A	68. J	93. E
19. B	44. B	69. C	94. I
20. A	45. F	70. E	95. B
21. A	46. I	71. B	96. A
22. J	47. H	72. E	97. F
23. H	48. C	73. I	98. C
24. G	49. E	74. C	99. H
25. B	50. J	75. A	100. G

Explanatory Answers

ITEM NUMBER	1	2	3	4	5	6	7	8	9	10
SYMBOL										
LETTER ANSWER	F	H	I	J	C	G	D	A	E	B
ITEM NUMBER	11	12	13	14	15	16	17	18	19	20
SYMBOL										
LETTER ANSWER	G	E	D	H	I	F	C	J	B	A
ITEM NUMBER	21	22	23	24	25	26	27	28	29	30
SYMBOL										
LETTER ANSWER	A	J	H	G	B	C	E	F	D	I

ITEM NUMBER	31	32	33	34	35	36	37	38	39	40
SYMBOL										
LETTER ANSWER	C	I	F	E	J	H	G	A	B	D
ITEM NUMBER	41	42	43	44	45	46	47	48	49	50
SYMBOL										
LETTER ANSWER	D	G	A	B	F	I	H	C	E	J
ITEM NUMBER	51	52	53	54	55	56	57	58	59	60
SYMBOL										
LETTER ANSWER	E	F	J	I	C	H	A	D	G	B
ITEM NUMBER	61	62	63	64	65	66	67	68	69	70
SYMBOL										
LETTER ANSWER	A	D	H	F	G	B	I	J	C	E
ITEM NUMBER	71	72	73	74	75	76	77	78	79	80
SYMBOL										
LETTER ANSWER	B	E	I	C	A	J	H	G	D	F
ITEM NUMBER	81	82	83	84	85	86	87	88	89	90
SYMBOL										
LETTER ANSWER	I	C	G	J	D	E	B	A	F	H
ITEM NUMBER	91	92	93	94	95	96	97	98	99	100
SYMBOL										
LETTER ANSWER	J	D	E	I	B	A	F	C	H	G

Counting Crosses and Zeros—Test II

Correct Answers

1. B	7. A	13. B	19. D
2. A	8. B	14. C	20. C
3. D	9. C	15. A	21. E
4. E	10. D	16. C	22. A
5. B	11. A	17. B	23. D
6. E	12. E	18. B	24. B

Explanatory Answers

Legend				Column A	Column B	Column C	Column D	Column E
	Number of Xs		Row 1	14 / 16	16 / 14	16 / 14	11 / 19	16 / 16
		Number of Os	Row 2	14 / 16	15 / 12	14 / 11	11 / 14	10 / 15
			Row 3	15 / 15	15 / 11	14 / 14	13 / 12	13 / 13
			Row 4	11 / 18	15 / 10	13 / 14	12 / 13	15 / 12

Examining Objects—Test III

Correct Answers

1. B	10. E	19. E	28. C
2. C	11. B	20. D	29. D
3. C	12. D	21. A	30. C
4. D	13. B	22. C	31. B
5. B	14. C	23. B	32. C
6. C	15. E	24. D	33. E
7. B	16. C	25. A	34. D
8. B	17. C	26. B	35. B
9. C	18. B	27. E	36. B

Explanatory Answers

1.	B	Bashed in.
2.	C	Two rounds—same numbers.
3.	C	Two squares—same numbers.
4.	D	Different numbers.
5.	B	Bashed in.
6.	C	Two squares—same numbers.
7.	B	Bashed in.
8.	B	Bashed in.
9.	C	Two squares—same numbers.
10.	E	Number space empty.
11.	B	Bashed in.
12.	D	Different numbers.
13.	B	Bashed in.
14.	C	Two rounds—same numbers.
15.	E	Number space empty.
16.	C	Two rounds—same numbers.
17.	C	Two squares—same numbers.
18.	B	Bashed in.
19.	E	Number space empty.
20.	D	Different numbers.
21.	A	All right.
22.	C	Two squares—same numbers.
23.	B	Bashed in.
24.	D	Different numbers.
25.	A	All right.
26.	B	Bashed in.
27.	E	Number space empty.
28.	C	Two rounds—same numbers.
29.	D	Different numbers.
30.	C	Two squares—same numbers.
31.	B	Bashed in.
32.	C	Two rounds—same numbers.
33.	E	Number space empty.
34.	D	Different numbers.
35.	B	Bashed in.
36.	B	Bashed in.

Mazes and Pursuits—Test IV

Correct Answers

1. D	11. E	21. G	31. G	41. D
2. A	12. F	22. A	32. E	42. F
3. B	13. G	23. F	33. D	43. B
4. F	14. H	24. B	34. B	44. E
5. H	15. D	25. F	35. A	45. A
6. G	16. B	26. A	36. C	46. C
7. C	17. E	27. C	37. H	47. G
8. E	18. D	28. B	38. E	48. H
9. C	19. C	29. D	39. F	49. E
10. A	20. H	30. H	40. G	50. C

Explanatory Answers

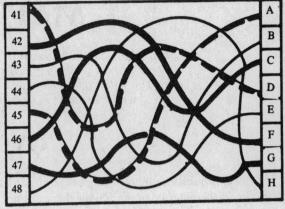

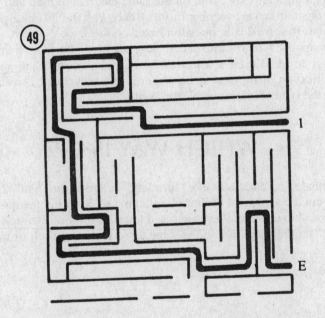

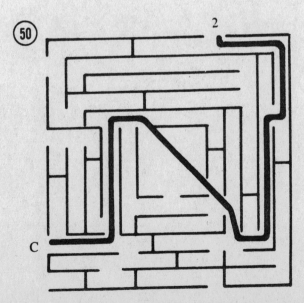

MAP SKILLS

Map skills questions measure your ability to keep a clear idea of where you are going in relation to the space in which you happen to be. You will be given clues and instructions to orient yourself within the map structure in order to navigate to a new destination.

OVERVIEW

Map skills questions tend to emphasize either where you are in a diagram or how to go from one spot to another on the diagram or map. Since these are not memorization questions, you will be allowed to use your pencil to write on the diagrams or maps as a way of testing your answer choices. When using your pencil to write on a diagram, be sure to write lightly. Erase any of your jottings that do not work out or that are no longer needed. If several questions are based on the same diagram or map and you make pencil markings for each question, the diagram can get very confusing. Hence, you should erase your markings for an earlier question before you start to solve a later question based on the same diagram.

Many diagrams or maps use symbols, also called "legends," to indicate such things as scale and direction. Look at the whole page to see if there is a key to the legend. A dotted line may indicate movement. An arrow may indicate direction of movement or of permitted movement. An important feature of many diagrams and maps is the symbol indicating North, South, East and West.

WHICH WAY IS UP?

Often questions are based on directions like "turn left," or "go to the right." The test-maker approaches the diagram or map from the viewpoint of the action taking place within the question. These directions do not necessarily correspond to the test-taker's right and left. The solution is simple. Turn the diagram or map so that "left" or "right" on the diagram is in the same direction as your left or right hand.

MAP READING TEST

DIRECTIONS: Answer question 1 on the basis of the street map and the information below.

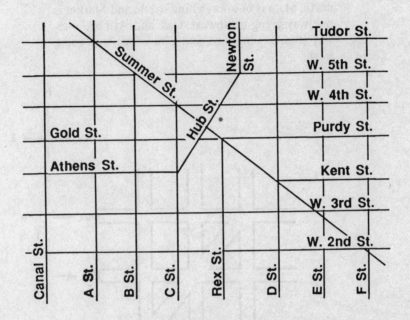

The city street map above shows an area that is divided into four sectors as follows:

Sector Adam: Bounded by Tudor Street, Newton Street, Hub Street, Athens Street, and Canal Street.

Sector Boy: Bounded by Tudor Street, F Street, West 4th Street, Hub Street, and Newton Street.

Sector Charles: Bounded by West 4th Street, F Street, West 2nd Street, C Street, and Hub Street.

Sector David: Bounded by Athens Street, C Street, West 2nd Street, and Canal Street.

1. Your house is in the block bounded by West 4th Street, Summer Street, and Hub Street. Your house is in Sector
   (A) David
   (B) Charles
   (C) Boy
   (D) Adam

Answer question 2 on the basis of the passage and diagram below.

As indicated by arrows on the street map shown below, Adams and River Streets are one-way going north. Main is one-way going south, and Market is one-way going northwest. Oak and Ash are one-way streets going east, and Elm is one-way going west.

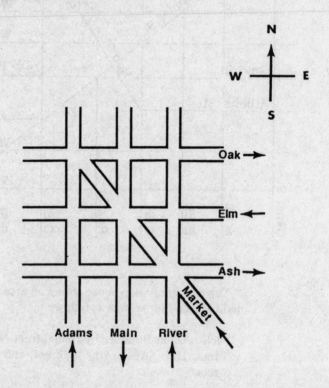

2. A taxi heading north on River Street between Ash and Elm Streets receives a call to proceed to the intersection of Adams and Oak. In order to travel the shortest distance and not break any traffic regulations, the taxi should turn
   (A) left on Elm and right on Market
   (B) left on Market and proceed directly to Oak
   (C) left on Elm and right on Adams
   (D) left on Oak and proceed directly to Adams

Answer questions 3 to 5 on the basis of the map below. The flow of traffic is indicated by the arrows. You must follow the flow of the traffic.

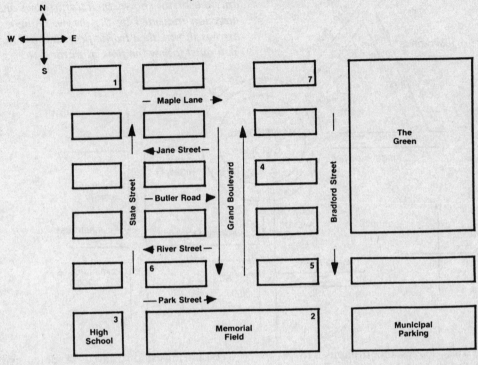

3. If you are located at point (1) and travel east one block, then turn right and travel for four blocks, and then turn left and travel one block, you will be closest to point
   (A) 6
   (B) 5
   (C) 3
   (D) 2

4. You are traveling by car from the corner of Butler Road and Bradford Street to Butler Road and Grand Boulevard. Which of the following is the most direct route?
   (A) Go one block south on Bradford Street, then go one block west on River Street, then go one block north on Grand Boulevard.
   (B) Go one block south on Bradford Street, then go two blocks west on River Street, then go one block north on State Street, and then go one block east on Butler Road.
   (C) Go one block north on Bradford Street, then go one block west on Jane Street, and then go one block south on Grand Boulevard.
   (D) Go two blocks south on Bradford Street, then go one block west on Park Street, then go two blocks north on Grand Boulevard.

5. You are driving from the high school to the corner of Maple Lane and Bradford Street. The best route to take would be to
   (A) go two blocks east on Park Street, then go four blocks north on Bradford Street
   (B) go two blocks north on State Street, then go two blocks east on Butler Road, then go two blocks north on Bradford Street
   (C) go four blocks north on State Street, then go two blocks east on Maple Lane
   (D) go one block east on Park Street, then go two blocks north on Grand Boulevard, then go one block east on Butler Road, then go two blocks north on Bradford Street

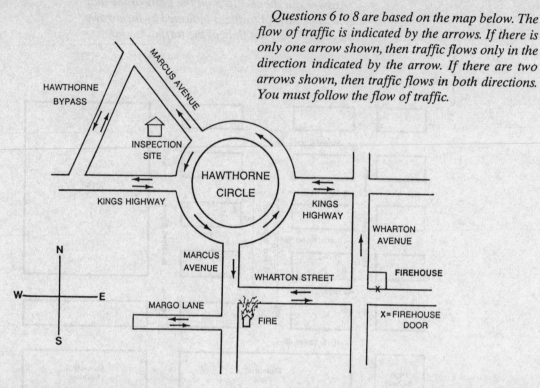

Questions 6 to 8 are based on the map below. The flow of traffic is indicated by the arrows. If there is only one arrow shown, then traffic flows only in the direction indicated by the arrow. If there are two arrows shown, then traffic flows in both directions. You must follow the flow of traffic.

6. A single fire truck leaves the firehouse to make a scheduled fire safety inspection of a building on the northern spur of Marcus Avenue. The best way to get there is to travel
   (A) north on Wharton Avenue, turn left onto Kings Highway, go clockwise around Hawthorne Circle, and exit the circle onto Marcus Avenue just north of Kings Highway
   (B) west on Wharton Street, turn right onto Marcus Avenue, enter Hawthorne Circle going counterclockwise, pass one Kings Highway exit, and exit onto Marcus Avenue
   (C) west on Wharton Street and immediately north on Wharton Avenue, turn left to go west on Kings Highway, turn right onto Hawthorne Circle to the next intersecting road, and turn right onto Marcus Avenue
   (D) north on Wharton Street, west on Kings Highway into Hawthorne Circle, then counterclockwise on Hawthorne Circle directly onto Marcus Avenue

7. While the inspection is in progress, the alarm sounds for a fire on the southern spur of Marcus Avenue. The firefighters interrupt their inspection, get on the truck, and proceed to the fire by traveling
   (A) southeast on Marcus Avenue, then turning right onto Hawthorne Circle, continuing past Kings Highway, and turning right onto Marcus Avenue

   (B) northwest on Marcus Avenue, making a sharp left turn onto the Hawthorne Bypass, making another left onto Kings Highway, then a right onto Hawthorne Circle, and the next right onto Marcus Avenue
   (C) northwest on Marcus Avenue, making a sharp left turn onto the Hawthorne Bypass, making a right onto Kings Highway, then a right onto Hawthorne Circle, and the next right onto Marcus Avenue
   (D) southeast on Marcus Avenue to Hawthorne Circle, clockwise around Hawthorne Circle, and past Kings Highway to the south exit for Marcus Avenue

8. Two other engines respond to this same fire described in question 7. Leaving directly from the fire station, these engines should travel
   (A) south on Wharton Street and turn left onto Marcus Avenue
   (B) north on Wharton Avenue, west on Kings Highway, counterclockwise on Hawthorne Circle past the northern spur of Marcus Avenue and the western extension of Kings Highway to the southern spur of Marcus Avenue
   (C) west on Wharton Street and south on Marcus Avenue
   (D) north on Wharton Avenue, west on Kings Highway, and clockwise around Hawthorne Circle to Marcus Avenue

Use the map below to answer questions 9 to 11. The flow of traffic is indicated by arrows. If there is only one arrow shown, then traffic flows only in the direction indicated by the arrow. If there are two arrows shown, then traffic flows in both directions. You must follow the flow of traffic.

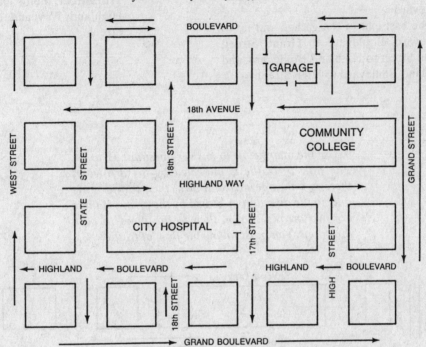

9. You are exiting the garage bound for a store located on the corner of State Street and Grand Boulevard. The quickest route to take is
   (A) out the east exit of the garage, north on High Street, south on Grand Street, west on Highland Boulevard to State Street, and south on State Street to the store
   (B) out the 18th Avenue exit of the garage and west to 18th Street, north on 18th Street to Boulevard, east on Boulevard to Grand Street, south on Grand Street to Highland Boulevard, west on Highland Way to State Street, and south on State Street to the store
   (C) out the west exit of the garage and up 17th Street to Boulevard, two blocks west on Boulevard to State Street, and all the way south on State Street to Grand Boulevard
   (D) out the south exit of the garage, west on 18th Avenue to State Street, then south on State Street to its corner with Grand Boulevard

10. As a result of an accident at the corner of State Street and Grand Boulevard, a victim must be transported by ambulance to City Hospital. The hospital emergency room entrance is located on 17th Street. Although the hospital is nearby, it is not easy to get to from the corner of State Street and Grand Boulevard. The best way to go is
    (A) Grand Boulevard to High Street, left onto High Street, left onto Highland Boulevard, and right onto 17th Street
    (B) Grand Boulevard to High Street, left onto High Street, right onto Highland Way, left onto Grand Street, left onto 18th Avenue, and left again onto 17th Street
    (C) east on Grand Boulevard to Highland Boulevard, west on Highland Boulevard to Highland Way, east on Highland Way, and south on 17th Street to the emergency room
    (D) east on Grand Boulevard to Grand Street, left onto Grand Street, then a left turn onto 18th Avenue, and south on 17th Street

11. You exit the garage bound for a building located on the southwest corner of High Street and Highland Boulevard. The best route to take is to go
(A) out the High Street exit of the garage and take High Street straight to Highland Boulevard
(B) out the east exit of the garage and up to Boulevard, Boulevard to Grand Street, Grand Street to Highland Boulevard, and Highland Boulevard to High Street
(C) out the west exit of the garage and straight down 17th Street to Highland Boulevard, then Highland Boulevard to its corner with High Street
(D) out the 18th Avenue exit of the garage to 17th Street, then south on 17th Street to Highland Way and the corner of High Street

Use the map below to answer questions 12 to 14. The flow of traffic is indicated by the arrows. If there is only one arrow shown, then traffic flows only in the direction indicated by the arrow. If there are two arrows shown, then traffic flows in both directions. You must follow the flow of traffic.

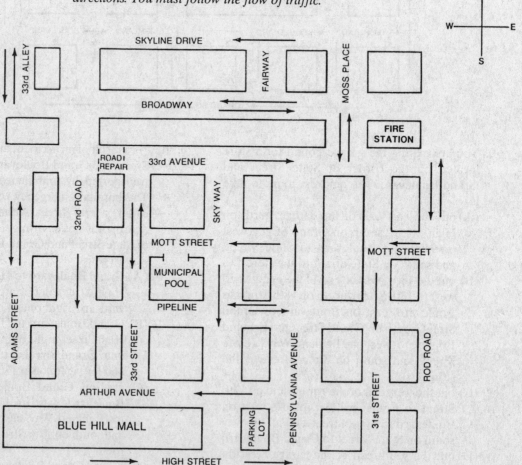

12. A patron at the municipal swimming pool had an accident. All emergency services crews are out on calls at this time, so the rescue team from the fire department must respond to the crisis. The best route for the rescue team to take from the fire station (which has exit doors at both east and west ends of the structure) is to

(A) exit by way of 31st Street, taking 33rd Avenue to Rod Road, making a right and taking Mott Street straight across to the pool

(B) exit the east end of the fire station, turn left onto Rod Road and then left onto Broadway, proceeding on Broadway to make a left onto 32nd Road, then turn right onto Mott Street to the pool

(C) exit onto Rod Road going north, then take Skyline Drive westbound to 32nd Road, go four blocks, turn left on Pipeline, then left onto Pennsylvania Avenue, and left from Pennsylvania Avenue onto Mott Street

(D) use the Moss Place exit of the fire station going north one-half block to Broadway, taking the long block of Broadway to a left onto 32nd Road and a right onto 33rd Avenue, followed by a turn south onto Sky Way and a right onto Mott Street

13. Two firefighters are performing a routine fire safety inspection at a store in the Blue Hill Mall. Their fire engine is parked just outside the Sky Way entrance to the parking lot. Their next assignment is to check a malfunctioning street box on the southeast corner of 32nd Road and 33rd Avenue. The most efficient way for them to get there is by

(A) going south on Sky Way, east on High Street to the corner of Rod Road, then left onto Rod Road as far as Broadway, west on Broadway, and south on 32nd Road to its corner with 33rd Avenue

(B) going around the parking lot to Arthur Avenue, then north onto 33rd Street and west onto 33rd Avenue to the corner

(C) going three-quarters of the way around the parking lot onto Arthur Avenue going west, then taking Arthur Avenue to make a right turn onto Adams Street northbound and a right onto 33rd Avenue

(D) continuing south on Sky Way to High Street, turning east for one block to make a left onto Pennsylvania Avenue, then right, left, and left again, so as to be headed west on Broadway, and left onto 32nd Road

14. Having repaired the street box, the firefighters are about to return to the fire station, but a road crew has just begun to repair a pothole on 33rd Avenue and has closed the street to all traffic between 33rd Street and 32nd Road. The firefighters will have to travel east on

(A) Broadway
(B) Pipeline
(C) Mott Street
(D) High Street

Answer question 15 on the basis of the diagram below. All streets are one way.

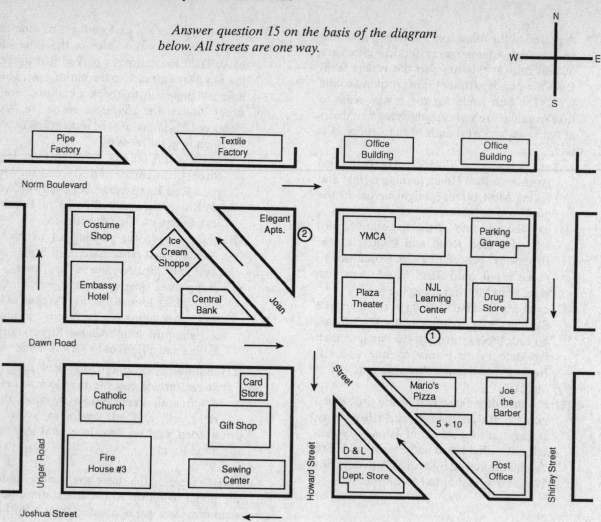

15. To drive from NJL Learning Center, located at (1), to Elegant Apartments, located at (2), via the shortest legal route, you would

(A) go east on Dawn Road, south onto Shirley Street, west onto Joshua Street, north on Unger Road, east on Norm Boulevard and south on Howard Street

(B) go east on Dawn Road, south on Shirley Street, west on Joshua Street, northwest on Joan Street, east on Norm Boulevard, and south on Howard Street

(C) go east on Dawn Road, south on Shirley Street, west on Dawn Road, and north on Howard Street

(D) go east on Dawn Road, right onto Shirley Street, east at Joshua Street, north at Unger Road, east on Dawn Road, and north on Howard Street

Map Reading Test

Correct Answers

1. D	4. A	7. B	10. D	13. C
2. A	5. C	8. C	11. B	14. B
3. D	6. C	9. D	12. D	15. B

Explanatory Answers

1. **(D)**

Sector Adam is bounded by heavy black lines.
The block with your house is marked with an X.

2. **(A)**

The route is marked on the diagram. Choice (C)
would be slightly longer. Choices (B) and (D)
are impossible from the given starting point.

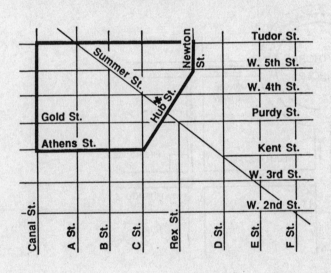

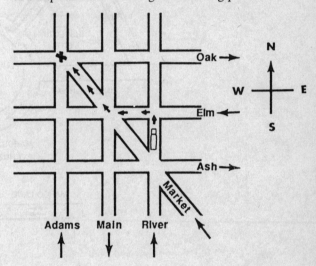

3. **(D)**

From point (1), east one block is right on Maple
Lane to Grand Boulevard; right four blocks is
south on Grand Boulevard to Park Street; left is
east into Park Street and one block brings us
closest to point (2).

4. **(A)**

Bradford Road is one-way southbound, so immediately eliminate choice (C). Park Street is one-
way eastbound, so eliminate choice (D). Choice
(B) is possible but is longer than choice (A), to
no advantage.

5. **(C)**

All other choices send you north on Bradford
Street.

6. (C)

The route is marked on the diagram. (A) is wrong because you cannot exit directly onto Wharton Avenue; you must go west on Wharton Street, then turn right onto Wharton Avenue. Even if you could exit onto Wharton Avenue, Hawthorne Circle moves counterclockwise, not clockwise. (B) sends you north on the spur of Marcus Avenue that is one-way southbound. As for (D), Wharton Street will never get you to Hawthorne Circle.

7. (B)

The route is marked on the diagram. This spur of Marcus Avenue is one-way in a northwesterly direction, so (A) is wrong. As for (C), the firefighters must make a right turn from Kings Highway onto Hawthorne Circle, not a left. (D) presents the same problem as does (A).

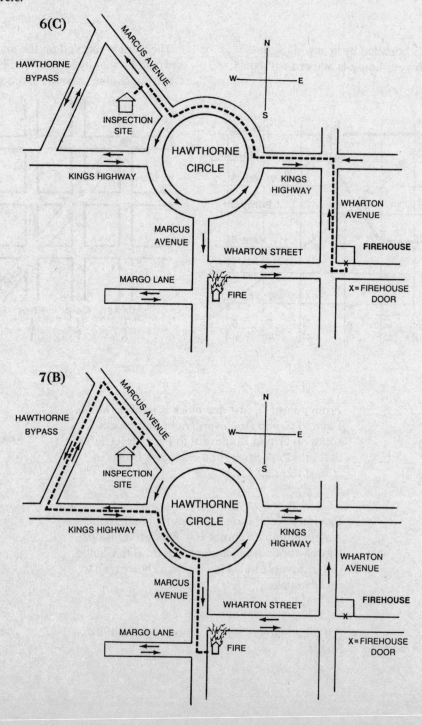

8. (C)

The route is marked on the diagram. (A) cannot be correct because Wharton Street runs east–west. (B) is a legal route, but it is unnecessarily long and roundabout. (D) travels the wrong way around Hawthorne Circle.

9. (D)

The route is marked on the diagram. (A) cannot be correct because High Street does not intersect Grand Street. (B) confuses Highland Boulevard with Highland Way. (C) goes the wrong way on 17th Street.

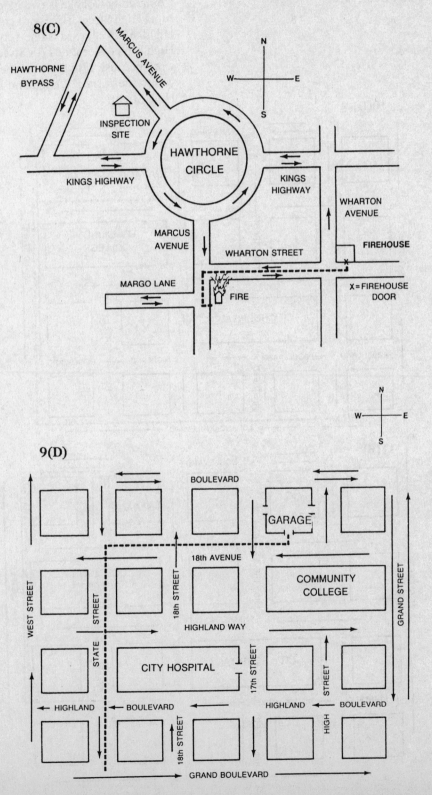

10. (D)

The route is marked on the diagram. Route (A) takes you the wrong way on the one-way 17th Street. (B) is a legal route, but a long one; the victim should get to the hospital as quickly as possible. As for (C), Grand Boulevard does not intersect Highland Boulevard.

11. (B)

See the diagram. (A) is impossible for two reasons: High Street is one-way northbound, and Highland Boulevard is south of the garage. Even if it were possible to go south on High Street, a straight run would be impossible because the Community College is two blocks wide and interrupts High Street. (C) is not correct because Highland Boulevard is one-way westbound, and High Street is east of 17th Street. (D) will not work because 17th Street alone will never get you to the corner of Highland Boulevard and High Street.

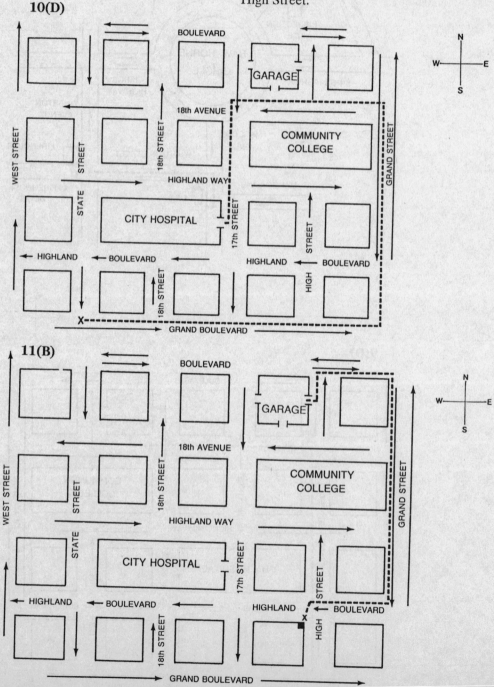

12. (D)

See the diagram. The answer cannot be (A) because there is no 31st Street exit and because Mott Street is not a through street. (B) ends up going east on a one-way westbound street. (C) would be legal and acceptable, but it is the long way round and so is not the answer to "Which is the best route?"

13. (C)

See the diagram. (A) is both possible and legal, but it is not the most efficient route. (B) is short and efficient, but it ends up going west on a one-way eastbound street. (D) is another possible route, but it is longer than (C) and involves many turns.

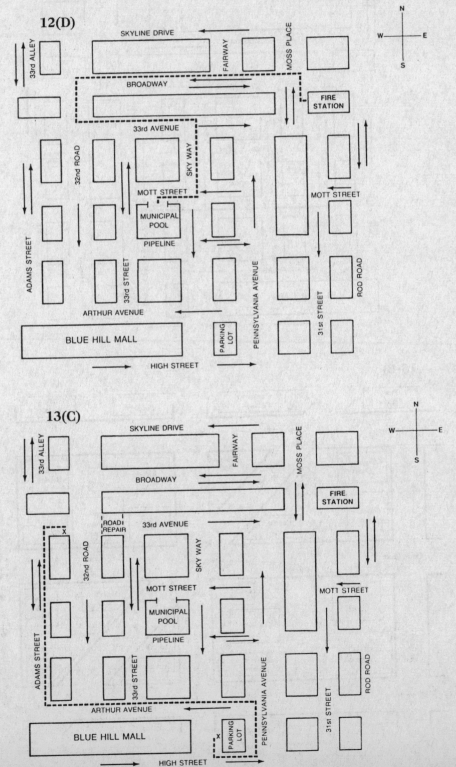

14. (B)

See the diagram. With 33rd Avenue closed, the firefighters have no way of getting to Broadway or, for that matter, to High Street from their current location. Mott Street is accessible, but they cannot go east on Mott Street.

15. (B)

See the diagram. (A) works, but it is a longer route than choice (B). As for (C), if you go west on Dawn Street after going south on Shirley Street, you are backtracking. In choice (D), if you go east at Joshua Street, you exit the map.

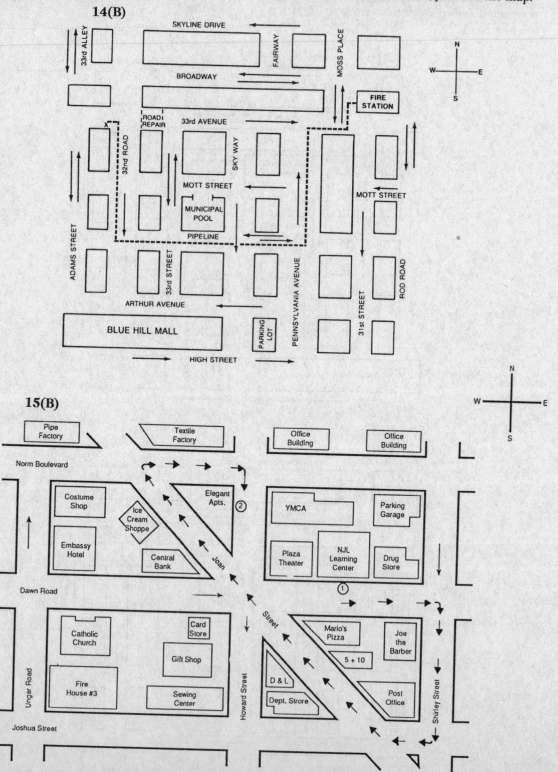

14(B)

15(B)

Part Two
Reasoning With Symbols

PART TWO

REASONING WITH SYMBOLS
Answer Sheets

To simulate actual examination conditions, mark your answers to each question in this chapter on these answer sheets. Make one clear, black mark for each answer. If you decide to change an answer, erase your error completely. On machine-scored examinations, additional or stray marks on your answer sheet may be counted as mistakes. After you have taken all the tests, compare your answers to the Correct Answers at the end of the chapter to see where you stand.

SYMBOL SERIES

Symbol Series—Test I

1 Ⓐ Ⓑ Ⓒ Ⓓ Ⓔ   10 Ⓐ Ⓑ Ⓒ Ⓓ Ⓔ   19 Ⓐ Ⓑ Ⓒ Ⓓ Ⓔ

2 Ⓐ Ⓑ Ⓒ Ⓓ Ⓔ   11 Ⓐ Ⓑ Ⓒ Ⓓ Ⓔ   20 Ⓐ Ⓑ Ⓒ Ⓓ Ⓔ

3 Ⓐ Ⓑ Ⓒ Ⓓ Ⓔ   12 Ⓐ Ⓑ Ⓒ Ⓓ Ⓔ   21 Ⓐ Ⓑ Ⓒ Ⓓ Ⓔ

4 Ⓐ Ⓑ Ⓒ Ⓓ Ⓔ   13 Ⓐ Ⓑ Ⓒ Ⓓ Ⓔ   22 Ⓐ Ⓑ Ⓒ Ⓓ Ⓔ

5 Ⓐ Ⓑ Ⓒ Ⓓ Ⓔ   14 Ⓐ Ⓑ Ⓒ Ⓓ Ⓔ   23 Ⓐ Ⓑ Ⓒ Ⓓ Ⓔ

6 Ⓐ Ⓑ Ⓒ Ⓓ Ⓔ   15 Ⓐ Ⓑ Ⓒ Ⓓ Ⓔ   24 Ⓐ Ⓑ Ⓒ Ⓓ Ⓔ

7 Ⓐ Ⓑ Ⓒ Ⓓ Ⓔ   16 Ⓐ Ⓑ Ⓒ Ⓓ Ⓔ   25 Ⓐ Ⓑ Ⓒ Ⓓ Ⓔ

8 Ⓐ Ⓑ Ⓒ Ⓓ Ⓔ   17 Ⓐ Ⓑ Ⓒ Ⓓ Ⓔ   26 Ⓐ Ⓑ Ⓒ Ⓓ Ⓔ

9 Ⓐ Ⓑ Ⓒ Ⓓ Ⓔ   18 Ⓐ Ⓑ Ⓒ Ⓓ Ⓔ   27 Ⓐ Ⓑ Ⓒ Ⓓ Ⓔ

SYMBOL ANALOGIES TEST
Symbol Analogies—Test I

1 Ⓐ Ⓑ Ⓒ Ⓓ Ⓔ   4 Ⓐ Ⓑ Ⓒ Ⓓ Ⓔ   7 Ⓐ Ⓑ Ⓒ Ⓓ Ⓔ   10 Ⓐ Ⓑ Ⓒ Ⓓ Ⓔ   13 Ⓐ Ⓑ Ⓒ Ⓓ Ⓔ

2 Ⓐ Ⓑ Ⓒ Ⓓ Ⓔ   5 Ⓐ Ⓑ Ⓒ Ⓓ Ⓔ   8 Ⓐ Ⓑ Ⓒ Ⓓ Ⓔ   11 Ⓐ Ⓑ Ⓒ Ⓓ Ⓔ   14 Ⓐ Ⓑ Ⓒ Ⓓ Ⓔ

3 Ⓐ Ⓑ Ⓒ Ⓓ Ⓔ   6 Ⓐ Ⓑ Ⓒ Ⓓ Ⓔ   9 Ⓐ Ⓑ Ⓒ Ⓓ Ⓔ   12 Ⓐ Ⓑ Ⓒ Ⓓ Ⓔ   15 Ⓐ Ⓑ Ⓒ Ⓓ Ⓔ

Symbol Analogies—Test II

1 Ⓐ Ⓑ Ⓒ Ⓓ Ⓔ	4 Ⓐ Ⓑ Ⓒ Ⓓ Ⓔ	7 Ⓐ Ⓑ Ⓒ Ⓓ Ⓔ	10 Ⓐ Ⓑ Ⓒ Ⓓ Ⓔ	13 Ⓐ Ⓑ Ⓒ Ⓓ Ⓔ
2 Ⓐ Ⓑ Ⓒ Ⓓ Ⓔ	5 Ⓐ Ⓑ Ⓒ Ⓓ Ⓔ	8 Ⓐ Ⓑ Ⓒ Ⓓ Ⓔ	11 Ⓐ Ⓑ Ⓒ Ⓓ Ⓔ	14 Ⓐ Ⓑ Ⓒ Ⓓ Ⓔ
3 Ⓐ Ⓑ Ⓒ Ⓓ Ⓔ	6 Ⓐ Ⓑ Ⓒ Ⓓ Ⓔ	9 Ⓐ Ⓑ Ⓒ Ⓓ Ⓔ	12 Ⓐ Ⓑ Ⓒ Ⓓ Ⓔ	15 Ⓐ Ⓑ Ⓒ Ⓓ Ⓔ

Symbol Analogies—Test III

1 Ⓐ Ⓑ Ⓒ Ⓓ Ⓔ	4 Ⓐ Ⓑ Ⓒ Ⓓ Ⓔ	7 Ⓐ Ⓑ Ⓒ Ⓓ Ⓔ	10 Ⓐ Ⓑ Ⓒ Ⓓ Ⓔ	13 Ⓐ Ⓑ Ⓒ Ⓓ Ⓔ
2 Ⓐ Ⓑ Ⓒ Ⓓ Ⓔ	5 Ⓐ Ⓑ Ⓒ Ⓓ Ⓔ	8 Ⓐ Ⓑ Ⓒ Ⓓ Ⓔ	11 Ⓐ Ⓑ Ⓒ Ⓓ Ⓔ	14 Ⓐ Ⓑ Ⓒ Ⓓ Ⓔ
3 Ⓐ Ⓑ Ⓒ Ⓓ Ⓔ	6 Ⓐ Ⓑ Ⓒ Ⓓ Ⓔ	9 Ⓐ Ⓑ Ⓒ Ⓓ Ⓔ	12 Ⓐ Ⓑ Ⓒ Ⓓ Ⓔ	15 Ⓐ Ⓑ Ⓒ Ⓓ Ⓔ

FIGURE CLASSIFICATION

Figure Classification—Test I

1 Ⓐ Ⓑ Ⓒ Ⓓ Ⓔ	6 Ⓐ Ⓑ Ⓒ Ⓓ Ⓔ	11 Ⓐ Ⓑ Ⓒ Ⓓ Ⓔ	16 Ⓐ Ⓑ Ⓒ Ⓓ Ⓔ	21 Ⓐ Ⓑ Ⓒ Ⓓ Ⓔ
2 Ⓐ Ⓑ Ⓒ Ⓓ Ⓔ	7 Ⓐ Ⓑ Ⓒ Ⓓ Ⓔ	12 Ⓐ Ⓑ Ⓒ Ⓓ Ⓔ	17 Ⓐ Ⓑ Ⓒ Ⓓ Ⓔ	22 Ⓐ Ⓑ Ⓒ Ⓓ Ⓔ
3 Ⓐ Ⓑ Ⓒ Ⓓ Ⓔ	8 Ⓐ Ⓑ Ⓒ Ⓓ Ⓔ	13 Ⓐ Ⓑ Ⓒ Ⓓ Ⓔ	18 Ⓐ Ⓑ Ⓒ Ⓓ Ⓔ	23 Ⓐ Ⓑ Ⓒ Ⓓ Ⓔ
4 Ⓐ Ⓑ Ⓒ Ⓓ Ⓔ	9 Ⓐ Ⓑ Ⓒ Ⓓ Ⓔ	14 Ⓐ Ⓑ Ⓒ Ⓓ Ⓔ	19 Ⓐ Ⓑ Ⓒ Ⓓ Ⓔ	24 Ⓐ Ⓑ Ⓒ Ⓓ Ⓔ
5 Ⓐ Ⓑ Ⓒ Ⓓ Ⓔ	10 Ⓐ Ⓑ Ⓒ Ⓓ Ⓔ	15 Ⓐ Ⓑ Ⓒ Ⓓ Ⓔ	20 Ⓐ Ⓑ Ⓒ Ⓓ Ⓔ	25 Ⓐ Ⓑ Ⓒ Ⓓ Ⓔ

Figure Classification—Test II

1 Ⓐ Ⓑ Ⓒ Ⓓ Ⓔ	6 Ⓐ Ⓑ Ⓒ Ⓓ Ⓔ	11 Ⓐ Ⓑ Ⓒ Ⓓ Ⓔ	16 Ⓐ Ⓑ Ⓒ Ⓓ Ⓔ	21 Ⓐ Ⓑ Ⓒ Ⓓ Ⓔ
2 Ⓐ Ⓑ Ⓒ Ⓓ Ⓔ	7 Ⓐ Ⓑ Ⓒ Ⓓ Ⓔ	12 Ⓐ Ⓑ Ⓒ Ⓓ Ⓔ	17 Ⓐ Ⓑ Ⓒ Ⓓ Ⓔ	22 Ⓐ Ⓑ Ⓒ Ⓓ Ⓔ
3 Ⓐ Ⓑ Ⓒ Ⓓ Ⓔ	8 Ⓐ Ⓑ Ⓒ Ⓓ Ⓔ	13 Ⓐ Ⓑ Ⓒ Ⓓ Ⓔ	18 Ⓐ Ⓑ Ⓒ Ⓓ Ⓔ	23 Ⓐ Ⓑ Ⓒ Ⓓ Ⓔ
4 Ⓐ Ⓑ Ⓒ Ⓓ Ⓔ	9 Ⓐ Ⓑ Ⓒ Ⓓ Ⓔ	14 Ⓐ Ⓑ Ⓒ Ⓓ Ⓔ	19 Ⓐ Ⓑ Ⓒ Ⓓ Ⓔ	24 Ⓐ Ⓑ Ⓒ Ⓓ Ⓔ
5 Ⓐ Ⓑ Ⓒ Ⓓ Ⓔ	10 Ⓐ Ⓑ Ⓒ Ⓓ Ⓔ	15 Ⓐ Ⓑ Ⓒ Ⓓ Ⓔ	20 Ⓐ Ⓑ Ⓒ Ⓓ Ⓔ	25 Ⓐ Ⓑ Ⓒ Ⓓ Ⓔ

SERIES REASONING

Number Series—Test I

1 Ⓐ Ⓑ Ⓒ Ⓓ Ⓔ   6 Ⓐ Ⓑ Ⓒ Ⓓ Ⓔ   11 Ⓐ Ⓑ Ⓒ Ⓓ Ⓔ   16 Ⓐ Ⓑ Ⓒ Ⓓ Ⓔ   21 Ⓐ Ⓑ Ⓒ Ⓓ Ⓔ

2 Ⓐ Ⓑ Ⓒ Ⓓ Ⓔ   7 Ⓐ Ⓑ Ⓒ Ⓓ Ⓔ   12 Ⓐ Ⓑ Ⓒ Ⓓ Ⓔ   17 Ⓐ Ⓑ Ⓒ Ⓓ Ⓔ   22 Ⓐ Ⓑ Ⓒ Ⓓ Ⓔ

3 Ⓐ Ⓑ Ⓒ Ⓓ Ⓔ   8 Ⓐ Ⓑ Ⓒ Ⓓ Ⓔ   13 Ⓐ Ⓑ Ⓒ Ⓓ Ⓔ   18 Ⓐ Ⓑ Ⓒ Ⓓ Ⓔ   23 Ⓐ Ⓑ Ⓒ Ⓓ Ⓔ

4 Ⓐ Ⓑ Ⓒ Ⓓ Ⓔ   9 Ⓐ Ⓑ Ⓒ Ⓓ Ⓔ   14 Ⓐ Ⓑ Ⓒ Ⓓ Ⓔ   19 Ⓐ Ⓑ Ⓒ Ⓓ Ⓔ   24 Ⓐ Ⓑ Ⓒ Ⓓ Ⓔ

5 Ⓐ Ⓑ Ⓒ Ⓓ Ⓔ   10 Ⓐ Ⓑ Ⓒ Ⓓ Ⓔ   15 Ⓐ Ⓑ Ⓒ Ⓓ Ⓔ   20 Ⓐ Ⓑ Ⓒ Ⓓ Ⓔ   25 Ⓐ Ⓑ Ⓒ Ⓓ Ⓔ

Letter Series—Test II

1 Ⓐ Ⓑ Ⓒ Ⓓ Ⓔ   4 Ⓐ Ⓑ Ⓒ Ⓓ Ⓔ   7 Ⓐ Ⓑ Ⓒ Ⓓ Ⓔ   10 Ⓐ Ⓑ Ⓒ Ⓓ Ⓔ   13 Ⓐ Ⓑ Ⓒ Ⓓ Ⓔ

2 Ⓐ Ⓑ Ⓒ Ⓓ Ⓔ   5 Ⓐ Ⓑ Ⓒ Ⓓ Ⓔ   8 Ⓐ Ⓑ Ⓒ Ⓓ Ⓔ   11 Ⓐ Ⓑ Ⓒ Ⓓ Ⓔ   14 Ⓐ Ⓑ Ⓒ Ⓓ Ⓔ

3 Ⓐ Ⓑ Ⓒ Ⓓ Ⓔ   6 Ⓐ Ⓑ Ⓒ Ⓓ Ⓔ   9 Ⓐ Ⓑ Ⓒ Ⓓ Ⓔ   12 Ⓐ Ⓑ Ⓒ Ⓓ Ⓔ   15 Ⓐ Ⓑ Ⓒ Ⓓ Ⓔ

SYMBOL SERIES

Symbols Series questions test your ability to spot the relationship governing a group of symbols so that you are able to choose the next term in the series. Each question consists of a series of five symbols on the left half of the page. Next to these are five other symbols labeled A, B, C, D and E. You are to study the first five symbols to determine what is happening in the series. Then select the one lettered symbol which best continues the series.

TWO SAMPLE QUESTIONS EXPLAINED

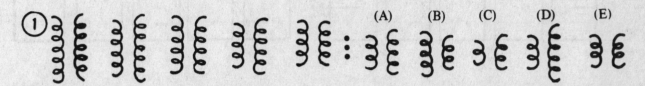

Each symbol in this series consists of two coils. The symbols differ from one another in the number of loops in each coil. In the first symbol, each coil has five loops; in the second, the left-hand coil has four loops and the right-hand coil has five loops; in the third, each coil has four loops. As this series pro-

gresses first the left-hand coil loses a loop and then the right-hand coil loses one. Since the fifth symbol in the series has three loops in each coil, the sixth symbol must have two loops in the left-hand coil and three loops in the right-hand coil as shown in the symbol labeled A.

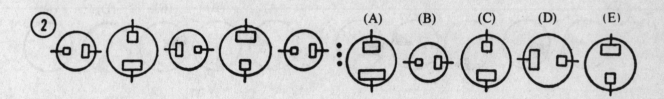

The first five symbols in question 2 show an alternation from small to large with a quarter-turn in a clockwise direction from one symbol to the next. Therefore, the next term must be a large circle (which eliminates alternative B) with the larger rectangle at the bottom of the circle (which eliminates alternatives

D and E). A closer look at alternative A shows that the rectangles within this circle are larger than the rectangles in any of the other circles. The best choice for the next term in this series is alternative C, which shows a large circle with a small square at the top and a larger rectangle on the bottom.

TEST QUESTIONS FOR PRACTICE

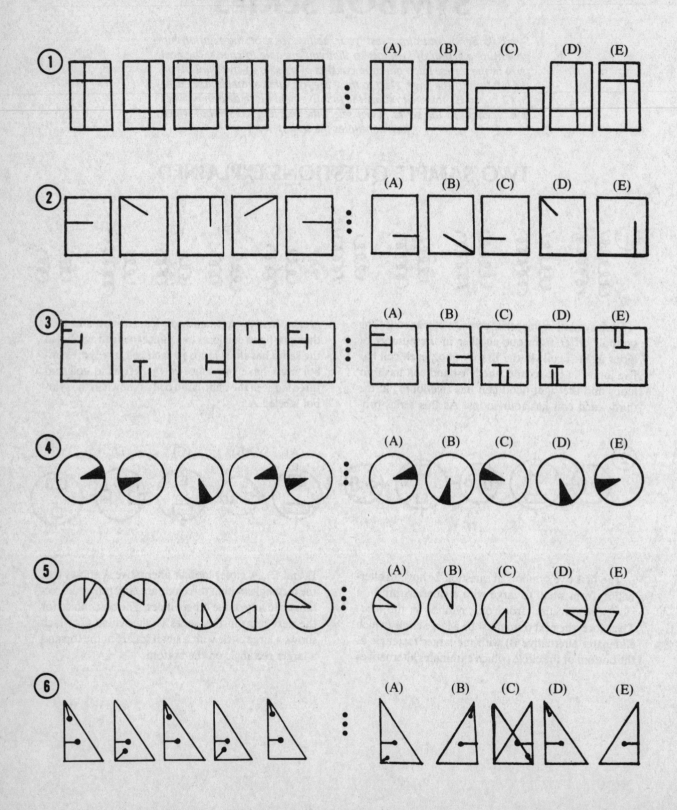

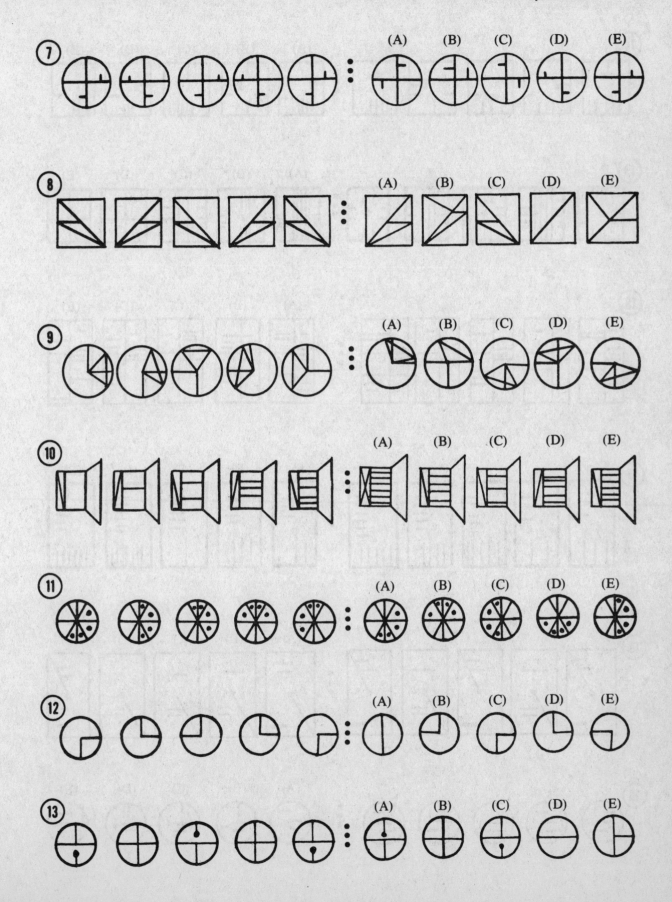

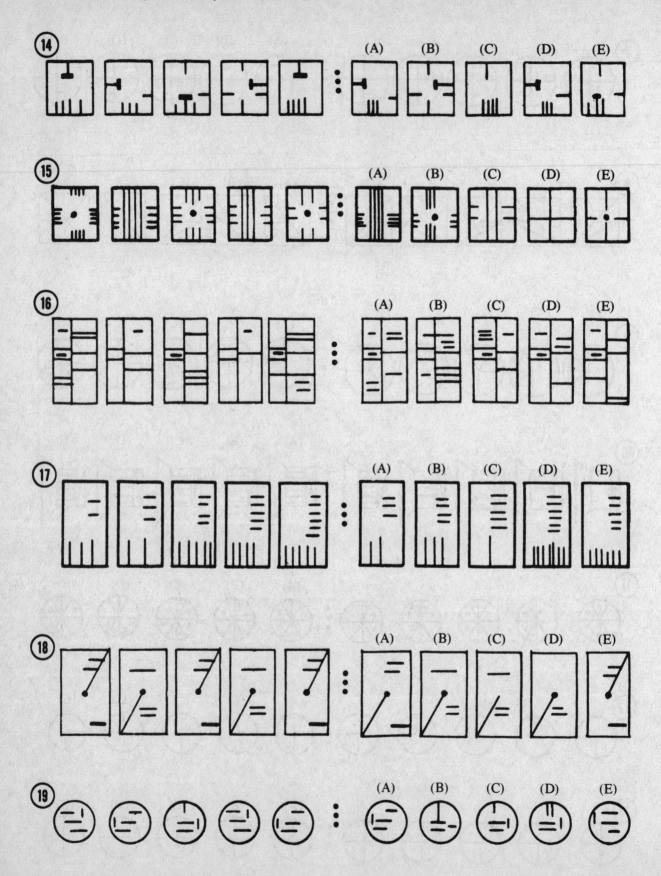

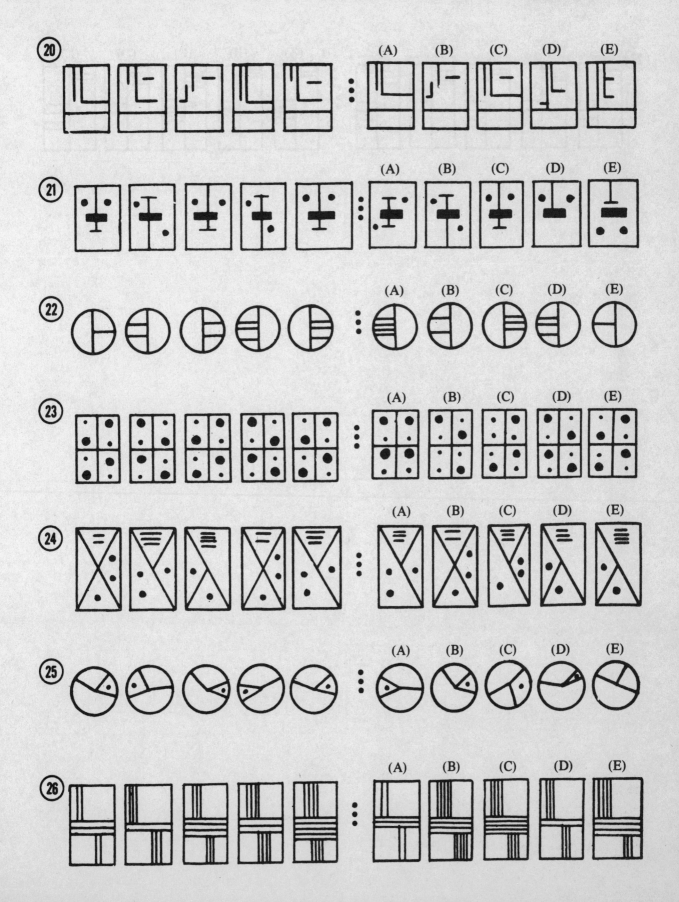

27

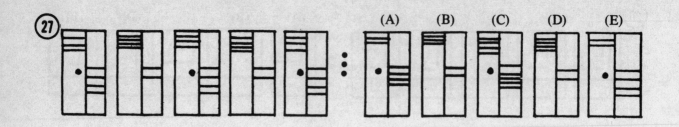

(A) (B) (C) (D) (E)

Symbol Series—Test I

Correct Answers

1. D	4. D	7. D	10. D	13. B	16. B	19. C	22. A	25. A
2. B	5. E	8. A	11. C	14. A	17. E	20. B	23. D	26. B
3. C	6. A	9. C	12. D	15. C	18. B	21. B	24. E	27. D

Explanatory Answers

1.	D	Horizontal line moves top-bottom-top . . . , vertical line moves left-right-left . . . .
2.	B	Line rotates clockwise ~ 45°.
3.	C	Symbols rotate counter-clockwise and interchange positions.
4.	D	Pattern repeats every third diagram.
5.	E	Angles alternate small-big-small . . . , the non-vertical or non-horizontal lines alternate sides about the vertical or horizontal respectively.
6.	A	Pattern repeats every second diagram.
7.	D	Pattern repeats every second diagram.
8.	A	Pattern repeats every second diagram.
9.	C	The quarter circle rotates 90° clockwise, the triangle rotates ~ 45° counter-clockwise.
10.	D	Slant line alternates \, /, \, / . . . , horizontal lines increase by 1.
11.	C	Dots rotate counter-clockwise 1 space.
12.	D	Odd position diagrams—the quarter circle alternates. Even position diagrams are all the same.
13.	B	Odd position diagrams—the dot alternates bottom-top-bottom. . . . Even position diagrams are all the same.
14.	A	Pattern repeats every fourth diagram.
15.	C	Notches decrease by 1 every second diagram. The dot and a decreasing number of vertical lines alternate.
16.	B	Odd position diagrams—the center left slot has a horizontal line. Even position diagrams—the center left slot is missing the horizontal line.
17.	E	2 horiz.-3 vert., 3 horiz.-2 vert., 4 horiz.-5 vert., 5 horiz.-4 vert., 6 horiz.-5 vert., 5 horiz.-6 vert.
18.	B	Pattern repeats every second diagram.
19.	C	Pattern repeats every third diagram.
20.	B	Pattern repeats every third diagram.
21.	B	Pattern repeats every second diagram.
22.	A	Horizontal lines alternate right-left-right . . . and increase by 1 after being used twice.
23.	D	Pattern repeats every second diagram.
24.	E	Pattern repeats every third diagram.
25.	A	Pattern repeats every second diagram.
26.	B	Two lines, two frames of 3 lines, two frames of 4 lines, 5 lines are next.
27.	D	Pattern repeats every second diagram.

SYMBOL ANALOGIES

Symbol Analogy questions are intended to measure your ability to find the underlying relationships that exist among groups of symbols. Each question consists of three boxes of symbols. The first box contains three symbols; the second box contains two symbols and a question mark; and the third box contains five symbols that are lettered A, B, C, D, and E. You must choose the one lettered symbol from the third box which can best be substituted for the question mark in the second box. In order to do this you must first discover what it is that the symbols in the first two boxes have in common and then further pinpoint how the that common feature varies between the first and second boxes. Your answer must then be that symbol which has the feature common to all the symbols in the series and yet maintains the same variation of that Symbol as exhibited by the other two symbols in the second set.

TWO SAMPLE QUESTIONS EXPLAINED

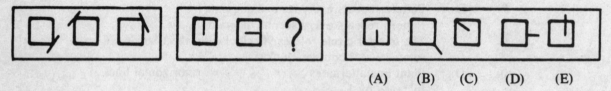

(A)   (B)   (C)   (D)   (E)

The common characteristic of this series is that each symbol consists of a square and a line that touches the square. The variation that distinguishes the symbols in set one from those in set two is the placement of the line in relation to the square. In the first set of symbols, the lines are tangent to one corner and outside each square. In the second set, the lines are perpendicular to one side and completely within each square. Only alternative A maintains this pattern.

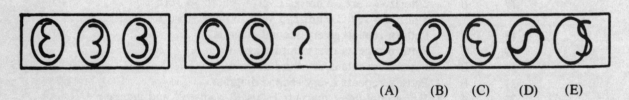

(A)   (B)   (C)   (D)   (E)

The common feature of this series is that all of the symbols are made up of similar curves enclosed within ovals. In the first set of symbols, two of the enclosed figures face in the same direction and one is reversed. Since the second set contains two enclosed figures facing in the same direction, the missing figure must be an oval which contains a figure that is the reverse of the two figures shown in the second set of symbols. Alternative B is the correct choice.

The following list will help you spot the common features and their variations
which are likely to turn up in Symbol Analogies questions.

COMMON FEATURES	VARIATIONS OF THE COMMON FEATURE
Lines that divide a figure	Equal or unequal divisions
Lines that form angles	Acute, obtuse, or right angles
Direction of lines	Vertical, horizontal, or slanted Pointed up or pointed down
Type of line	Solid or broken Curved or straight All the same, some different, or all different
Number of lines	2, 3, 4, etc. Same number in each figure or different number in each figure
Relationship among lines	Intersecting or non-intersecting Parallel or not parallel
Relationship of lines to figures	Lines inside or outside figures Lines touching or not touching figures
Closed shapes	Formed of straight lines or curved lines
Open figures	Open end up or down, left or right
Direction of figure	Pointed up or down, left or right
Shape of figure	Same shapes or different shapes
Shading of figure	Wholly or partially shaded Shaded with horizontal, vertical, or slanted lines
Size of figure	Small or large Same size or different sizes
Relationships between figures	Touching, separate Overlapping, sharing a common side
Figures within figures	Same figures or different figures Concentric figures or not concentric

TEST I. SYMBOL ANALOGIES

DIRECTIONS: Each question consists of two sets of symbols that are analogous to each other. That means the sets share a common characteristic while they differ in a specific aspect of that characteristic. In each question, the first set contains three symbols and the second set contains two symbols and a question mark. Following the symbol sets are five alternatives labeled A, B, C, D and E. You must choose the one lettered symbol which can best be substituted for the question mark. The correct choice will have the characteristic common to both sets of symbols and yet maintain the same variation of that characteristic as the two symbols in the second set. Correct and explanatory answers follow this text.

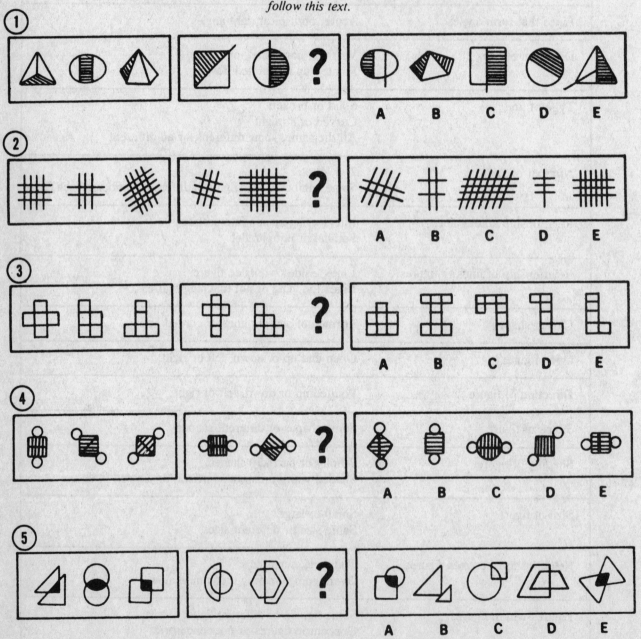

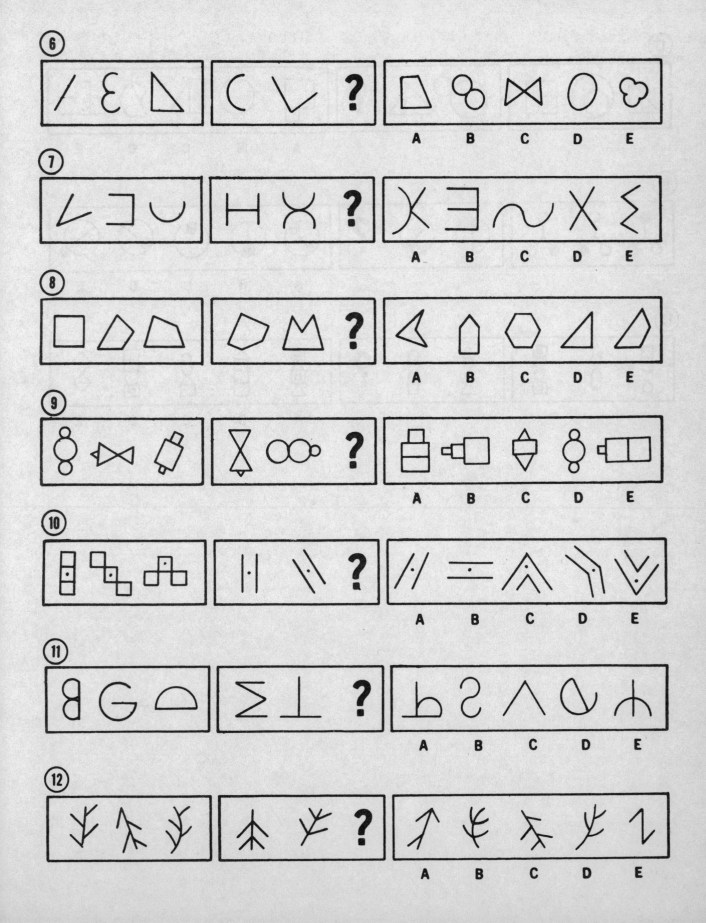

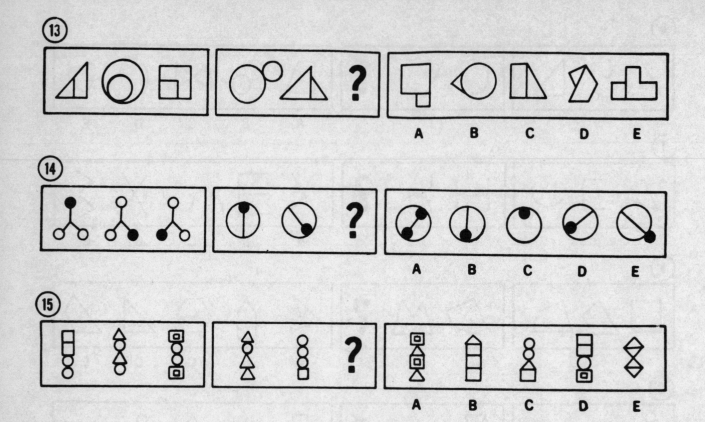

13

A B C D E

14

A B C D E

15

A B C D E

TEST II. SYMBOL ANALOGIES

DIRECTIONS: Each question consists of two sets of symbols that are analogous to each other. That means the sets share a common characteristic while they differ in a specific aspect of that characteristic. In each question, the first set contains three symbols and the second set contains two symbols and a question mark. Following the symbol sets are five alternatives labeled A, B, C, D and E. You must choose the one lettered symbol which can best be substituted for the question mark. The correct choice will have the characteristic common to both sets of symbols and yet maintain the same variation of that characteristic as the two symbols in the second set. Correct and explanatory answers follow this test.

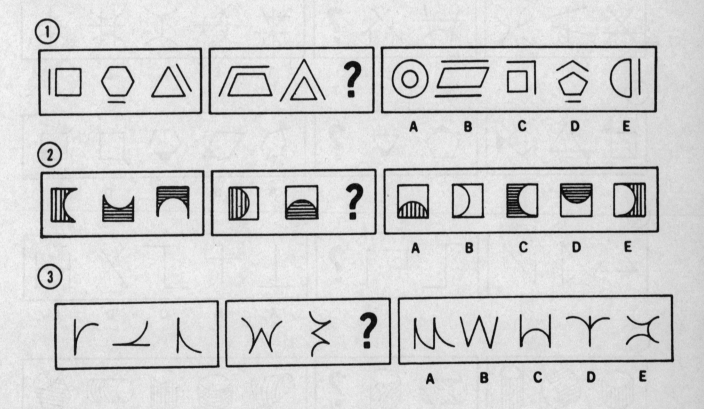

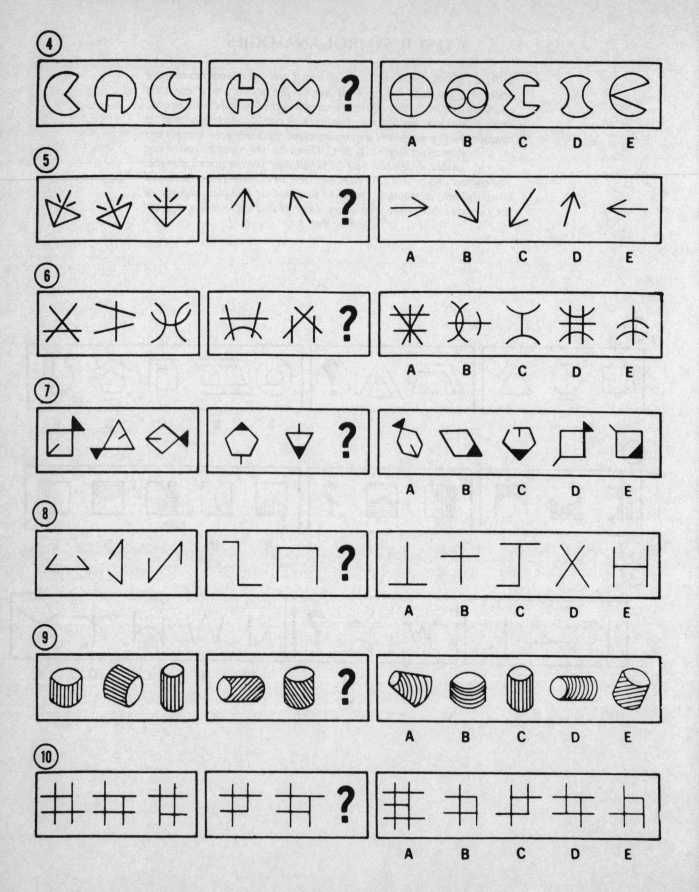

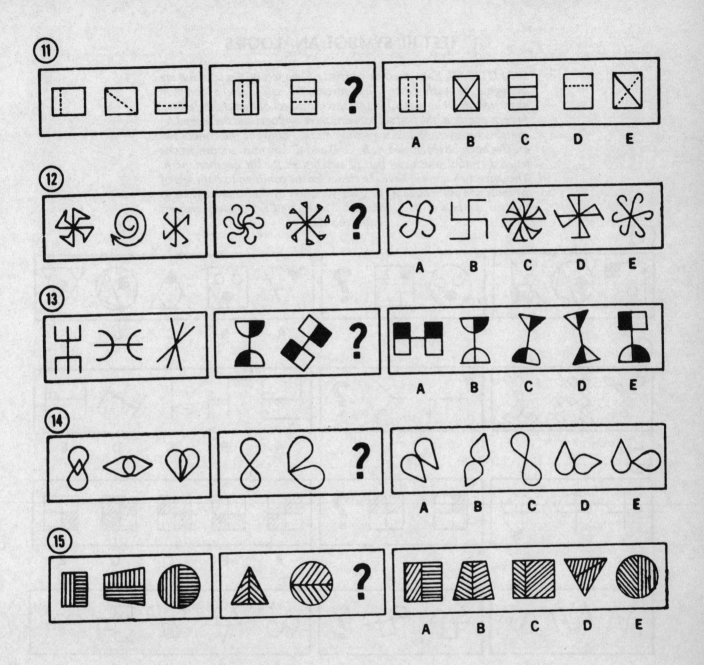

TEST III. SYMBOL ANALOGIES

DIRECTIONS: Each question consists of two sets of symbols that are analogous to each other. That means the sets share a common characteristic while they differ in a specific aspect of that characteristic. In each question, the first set contains three symbols and the second set contains two symbols and a question mark. Following the symbol sets are five alternatives labeled A, B, C, D and E. You must choose the one lettered symbol which can best be substituted for the question mark. The correct choice will have the characteristic common to both sets of symbols and yet maintain the same variation of that characteristic as the two symbols in the second set. Correct and explanatory answers follow this text.

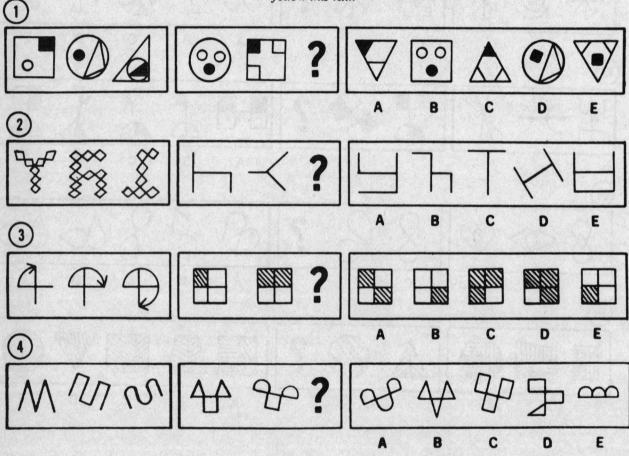

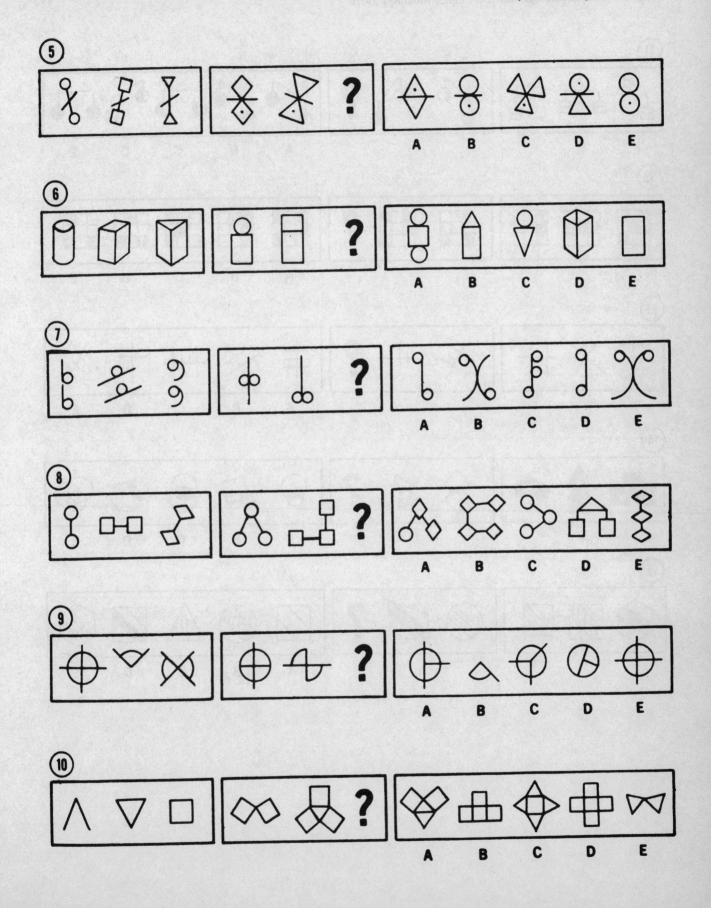

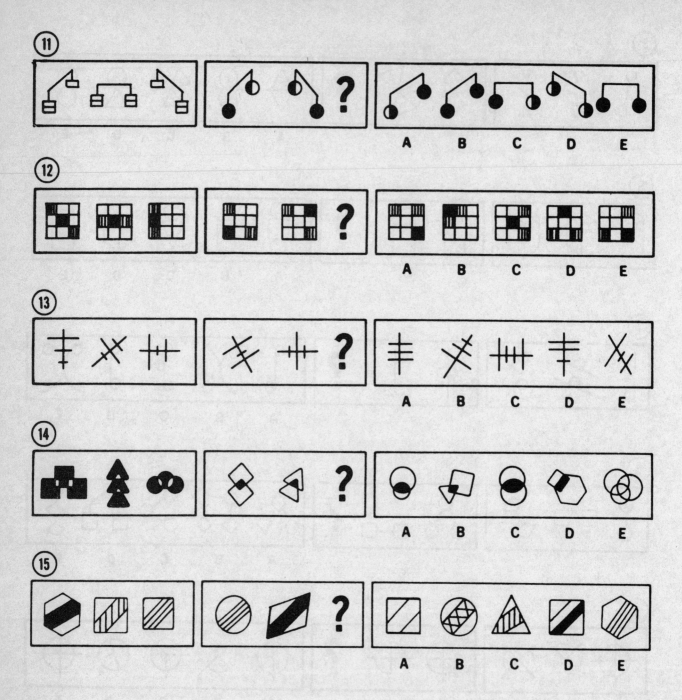

Symbol Analogies—Test I

Correct Answers

1. C	4. A	7. D	10. C	13. A
2. B	5. D	8. B	11. C	14. D
3. D	6. E	9. E	12. B	15. B

Explanatory Answers

1.   C   The similarity between the two sets is that each is made up of closed figures which contain a shaded portion. In the first set, the figures are divided into thirds and one-third of each figure is shaded. In the second set, the figures are divided in half with the dividing line extended beyond the perimeter of the figure and one-half of each figure is shaded. Only alternative C meets both of these requirements.

2.   B   Both sets of figures consist of lines that cross each other. In the first set, each figure has one more cross line in one direction than it does in the other. In the second set, each figure has an equal number of lines going in each direction. Alternative B is the only choice which has the same number of lines in each direction.

3.   D   The two sets are similar in that both consist of groups of boxes. They differ in that each figure in the first set is made of five boxes, while each figure in the second set is made up of six boxes. Therefore, the missing figure must also consist of six boxes as does figure D.

4.   A   Each figure in this series consists of a shaded rectangle bisected by a straight line and two circles that touch the outer edge of each rectangle. In the first set, the circles touch each rectangle at points opposite to the bisecting line. In the second set, the circles outside each rectangle are connected by the bisecting line. Alternative A is the only symbol that can complete the second set according to the pattern established.

5.   D   All the figures in this series are made up of overlapping shapes. The first set consists of identical shapes that overlap each other. In the second set, each figure is overlapped by one-half of a larger but similar shape. Figure D follows the pattern established by the first two figures in the set.

6.   E   All the figures in this series consist of straight or curved line segments. In the first set, the figures are made up of an increasing number of line segments with one segment in the first figure, two segments in the second figure, and three segments in the third figure. Likewise, in the second set the first figure has one segment and the second figure has two segments. The missing figure must, therefore, have three segments as does figure E.

7.   D   All the symbols in this series consist of straight or curved lines. Each figure in the second set consists of a repeat of a figure in the first set together with its opposite. The only figure from the

first set that is missing in the second set is the angle, which, when combined with its opposite, creates the figure represented by alternative D.

8.   B   Both sets consist of closed figures. The figures in the first set are all four-sided, while the figures in the second set are five-sided. The missing figure, therefore, must be B, which is the only five-sided figure offered.

9.   E   Each figure in this series is made up of three similar shapes. In the first set the three shapes in each figure vary in size from small to large to medium. In the second set, the three shapes that make up each figure vary in size from large to large to small. Only alternative E follows the pattern established by the two figures in the second set.

10.   C   The similarity between these sets is directional. The parallel lines in the second set repeat the direction of the boxes in the first set; therefore, the missing figure must be alternative C.

11.   C   The only thing these figures have in common is that all are composed of lines. The figures in the first set consist of both straight and curved lines. The figures in the second set are made up of only straight lines. Figure C is the only alternative which consists solely of straight lines.

12.   B   All of the symbols in this series consist of lines branching out from a stem. In the first set the lines branch out from alternating positions on either side of the stem. In the second set the lines branch out from points directly opposite each other on either side of the stem. Only in figure B are the branches directly opposite each other.

13.   A   All of the figures in this series consist of two similar shapes which adjoin. In the first set, the smaller shape is completely enclosed by the larger one. In the second set, the smaller shape lies outside the larger one. Only figure A has two identical shapes which are in the same relationship to each other as are the two given figures in the second set.

14.   D   Each figure in the first set is identical except for the position of the shaded ball. Similarly, each figure in the second set is identical with the shaded ball in a position that exactly corresponds to its position in the first set. Figure D, which shows the shaded ball in the lower left portion of the circle corresponds to the third figure in the first set in which the dark ball is also in the lower left portion of the figure.

15.   B   All of the figures in this series are composed of two different shapes. In the first set, each symbol is made up of two of each shape. In the second set each figure consists of one shape that appears three times and one shape that appears only once. Figure B, which consists of three rectangles and one triangle, is the only figure that conforms to the pattern established.

Symbol Analogies—Test II

Correct Answers

1. C	4. D	7. E	10. D	13. D
2. D	5. D	8. A	11. E	14. C
3. A	6. B	9. E	12. C	15. B

Explanatory Answers

1.   C   All of the symbols in this series consist of a closed shape and lines outside and parallel to the shape. In the first set, there is one line outside each shape and parallel to one side. In the second set, there are two adjoining lines outside each closed shape. These outside lines run parallel to two of the sides of each closed shape. Only figure C follows this pattern.

2.   D   Each figure consists of both straight and curved lines, and each is shaded. In the first set, the lines in the shaded area run parallel to the longest straight edge. In the second set, the lines in the shaded area run parallel to the diameter of the semi-circle. Only in alternative D is this pattern maintained.

3.   A   Each symbol in this series consists of two or more lines. In the first set, each figure has two lines — one straight and one curved. In the second set, each figure has four lines — two straight and two curved. Figure A is the only symbol that can complete the second set.

4.   D   All of the figures in this series consist of circles with some part cut out. In the first set there is first a triangular, then a square and then a curved cut-out in the circles. In the second set, the cut-outs follow the same pattern, but here there are two cut-outs in each circle. Since the triangular and square cut-outs are shown, the missing figure in the second set must be a circle with two curved cut-outs as in alternative D.

5.   D   All of the figures in this series are essentially arrows, but they differ in the direction in which they point. All of the figures in the first set are pointing down and all of the figures in the second set are pointing up. Therefore, the missing arrow must be alternative D, the only one pointing in the right direction.

6.   B   All of the figures in this series are made up of lines, either straight or curved, or a combination of the two. However, each figure in the first set consists of three lines, while each figure in the second set consists of four lines. Only alternative B presents a figure made up of four lines, which is what is needed to complete the second set.

7.   E   Each symbol in this series consists of a closed shape, a small dark triangle and a short line. In the first set, the small dark triangle is outside the closed shape with the short line opposite and inside the figure. In the second set, the small dark triangle is inside each figure and the short line opposite and outside. Alternative E, which meets both these conditions, is the figure needed to complete the second set.

8.   A   All of the figures in this series consist of straight lines that form angles. In the first set, each group of lines forms two acute angles. In the second set, each group of lines forms two right angles. The only alternative which forms two right angles is A.

9.   E   The common feature of this series is that each symbol is a cylinder with lines on the outside surface. In the first set, the lines all run perpendicular to the base. In the second set, the lines all run diagonal to the base. Alternative E is the only one that correctly completes the set.

10.   D   Each figure in this series consists of four lines — some long and some short. All of the figures in the first set are made up of three long lines and one short line. All of the figures in the second set consist of two long lines and two short ones. Therefore, alternative D is the one that is needed to complete the set.

11.   E   The common feature of this series is that each symbol is a rectangle with lines inside it. In the first set, each rectangle has one broken line within it. In the second set, each rectangle contains one broken and one solid line. Such a figure is illustrated by alternative E.

12.   C   Each figure in this series represents rotation. In the first set, the figures indicate rotation in a clockwise direction, while in the second set, the figures indicate counterclockwise rotation. Figure C, which represents counterclockwise rotation, is the one that is needed to complete the set.

13.   D   The symbols in the first set are repeated in the second set except that in the second set, the ends are closed and opposite segments are shaded. Since the first two symbols from the first set already appear in the second set, it must be a variation of the third symbol which is needed to complete the set. Such a figure is provided by alternative D.

14.   C   All of the symbols in this series are teardrops. In the first set, the teardrops overlap. In the second set, the pointed ends touch. The missing figure must, therefore, be C which consists of teardrop shapes with points touching.

15.   B   The common feature of these sets is that each symbol is a closed shape with lines in it. In the first set, the enclosed lines run at right angles to each other. In the second set, each shape is divided in half and the enclosed lines form a herringbone pattern as in alternative B.

Symbol Analogies—Test III

Correct Answers

1. C	4. C	7. E	10. D	13. D
2. A	5. B	8. E	11. E	14. C
3. D	6. B	9. B	12. E	15. C

Explanatory Answers

1. **C** All of the symbols in this series consist of large closed shapes which contain a number of smaller shapes. In the first set, each large shape encloses one similar shape, which is darkened, and one different shape, which is unshaded. In the second set, each large shape encloses three similar shapes one of which is darkened. Only alternative C maintains this pattern.

2. **A** The common feature of this series is the shape of the symbols. The shapes in the first set, which are formed by a series of small diamonds, are repeated in the second set with simple straight lines. Since the first and last shapes from the first set already appear in the second set, the missing figure must be analogous to the middle figure of the first set. Alternative A is such a shape.

3. **D** The symbols in this series are growing longer in a clockwise direction. The first set progresses from one-quarter to one-half to three-quarters of a circle. The second set shows first one-quarter of a square darkened, then one-half of a square darkened. The missing figure must, therefore, be three-quarters of a square darkened as in alternative D. Alternative C is incorrect because it does not allow the darkened segments to progress in a clockwise direction.

4. **C** Each symbol in this series is made up of three elements with the midsection of each figure headed in the opposite direction from the two outer sections of each figure. In the first set, each figure consists of three identical elements only the center portion of each symbol is reversed. The second set repeats the shapes used in the first set for the upper half of each figure, but substitutes a square for each midsection. Alternative C is the only one that follows the pattern established by this series.

5. **B** Each figure in this series consists of two similar shapes and a line separating them. In the first set, the similar shapes are connected by a straight line and separated by a slanted line. In the second set, the similar shapes are separated by a horizontal line and in each case, the lower shape contains a dot. Only alternative B maintains this pattern.

6. **B** This series presents figures shown first in perspective and then in a single dimension. The first set pictures a cylinder, a box and a prism in perspective. The second set pictures a single-dimensional rendition of two different faces of each of the figures in the first set. Since two faces of the cylinder and the box appear in the second set, the missing figure must be two faces of the prism as in alternative B.

7. **E** The common feature of this series is that all of the symbols consist of lines with small circles attached at various points along the line. The variation is that in the first set, each figure consists of two lines and circles faced in the same direction while in the second set each figure consists of the same two lines and circles back to back. The missing figure is, therefore, the third figure in the first set with the two lines and circles back to back as in alternative E.

8. **E** The symbols in this series each consist of similar closed shapes joined by short lines. In the first set, each symbol consists of

two identical shapes joined by a line, while in the second set each symbol consists of three identical shapes joined by lines. Only alternative E continues the pattern of three identical shapes joined by short lines.

9.   B   All the symbols in this series are parts of a circle bisected by straight lines. In the first set, each bisecting line extends beyond the circle, while in the second set only two of the bisecting lines extend beyond the circle. Alternative B, which pictures one-quarter of the circle with one line extending beyond the circle and one line contained within the circle is the figure needed to complete the set.

10.   D   Each symbol in this series consists of a figure made up of straight lines. The figures in the second set repeat the figures in the first set, building a square off each side of the original figure. Since the angle and the triangle of the first set already appear in the second set, the missing figure must be based on the square as is alternative D.

11.   E   The symbols in this series consist of identical arrangements of similar shapes of varying sizes connected by armatures. In the first set, the connected shapes are either double or single rectangles and in the second set the connected shapes are either all dark or dark and light circles. In each figure, however, the all dark circles correspond to the double rectangles and the dark and light circles correspond to the single rectangles. Since the first and last figures from the first set already appear in the second set, the missing figure must be the arrangement analogous to the middle figure of the first set. This is represented by alternative E.

12.   E   The common feature of this series is that each symbol consist of a grid with one dark square and two shaded squares. In the first set, the dark square is centered between two shaded squares and each shaded square touches the dark square. In the second set, the dark square is in one corner with the two shaded squares in adjacent corners. Only alternative E maintains this pattern.

13.   D   This series consists of figures made up of a single line crossed by three lines. In the first set the three cross lines are each a different length. In the second set two of the cross lines are the same length and one is a different length. This configuration is represented by alternative D.

14.   C   The symbols in this series each consists of identical shapes that overlap each other. In the first set, three identical dark shapes overlap and the overlapping area is white. In the second set two identical white shapes overlap and the overlapping area is dark. Alternative C maintains this pattern.

15.   C   In this series each figure consists of a closed shape with a pattern of lines inside it. Since all the closed shapes are different, they need not be considered. This leaves only the lines inside each shape to determine the relationship. In the first set there is a solid line, a series of widely spaced vertical lines and a series of closely spaced slanting lines. The closely spaced slanting lines and the solid lines are repeated in the second set, so the missing figure must be a closed shape that encloses widely spaced vertical lines as does alternative C.

FIGURE CLASSIFICATION

In Figure Classification Tests, each problem consists of two groups of figures which are labeled 1 and 2. These two groups are followed by five answer figures which are lettered A, B, C, D and E. For each problem you must decide what characteristic is common to EACH of the figures in group 1 but appears in NONE of the figures in group 2. Then select the lettered answer figure that has this characteristic. The sample problems that follow should make the directions quite clear.

THREE SAMPLE QUESTIONS EXPLAINED

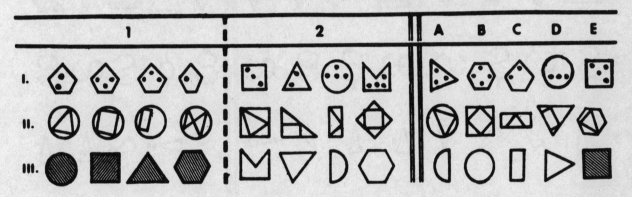

In sample problem I, all the figures in group 1 are pentagons, but none of the figures in group 2 is a pentagon; therefore, C, the only pentagon among the answer figures, is the correct choice.

In sample problem II, all the figures in group 1 include a circle, but none of the figures in group 2 includes a circle; so A is the correct answer.

In sample problem III, all the figures in group 1 are shaded, but none of the figures in group 2 is a shaded figure; therefore, the correct answer is E, the only shaded answer figure.

TEST I. FIGURE CLASSIFICATION

DIRECTIONS: In this type of test, each problem consists of two groups of figures labeled 1 and 2. These two groups are followed by five answer figures, lettered A, B, C, D, and E. For each problem you must decide what characteristic each of the figures in Group 1 has that none of the figures in Group 2 has. Then select the lettered answer figure that has this characteristic. Correct and explanatory answers follow this test.

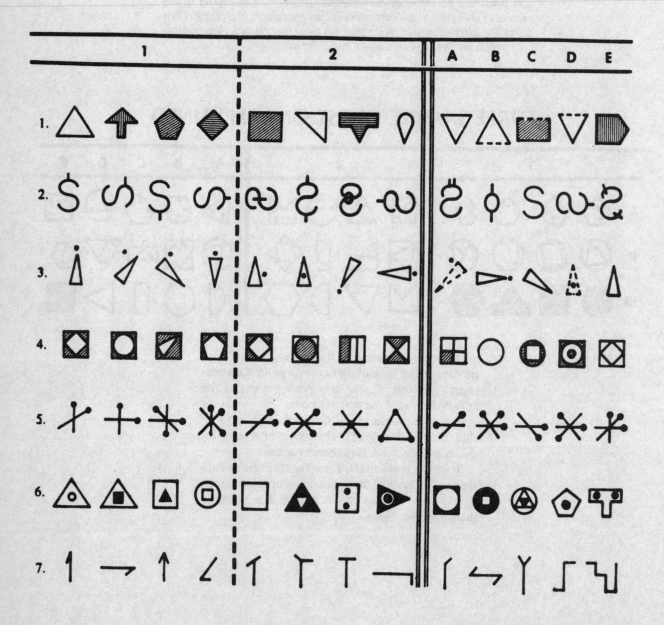

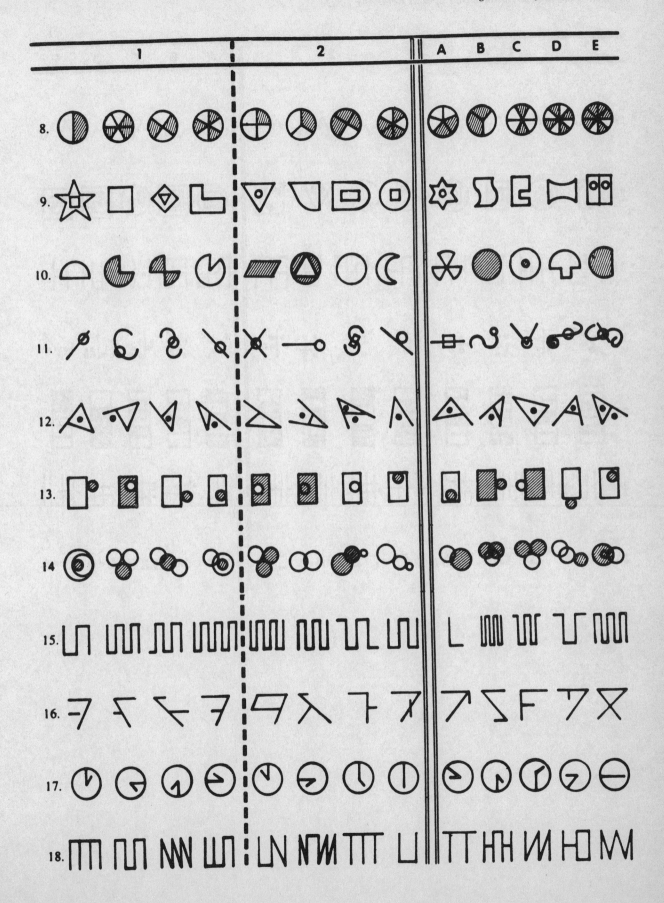

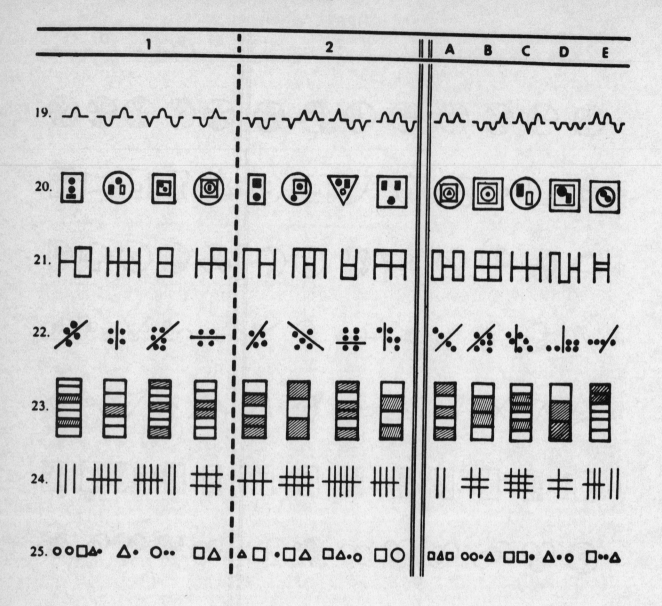

TEST II. FIGURE CLASSIFICATION

DIRECTIONS: Each of these problems consists of two groups of figures, labeled 1 and 2. These are followed by five lettered answer figures. For each problem you are to decide what characteristic each of the figures in group 1 has that none of the figures in group 2 has. Then select the lettered answer figure that has this characteristic. Correct and explanatory answers follow this test.

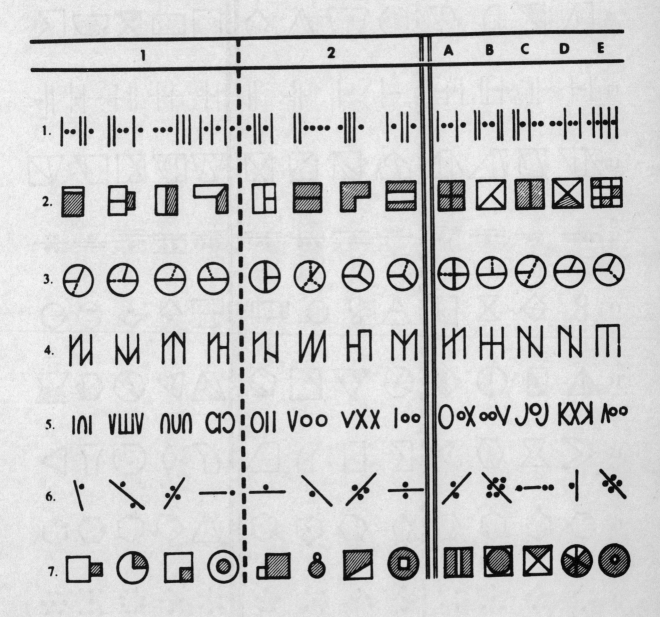

| | 1 | 2 | A B C D E |

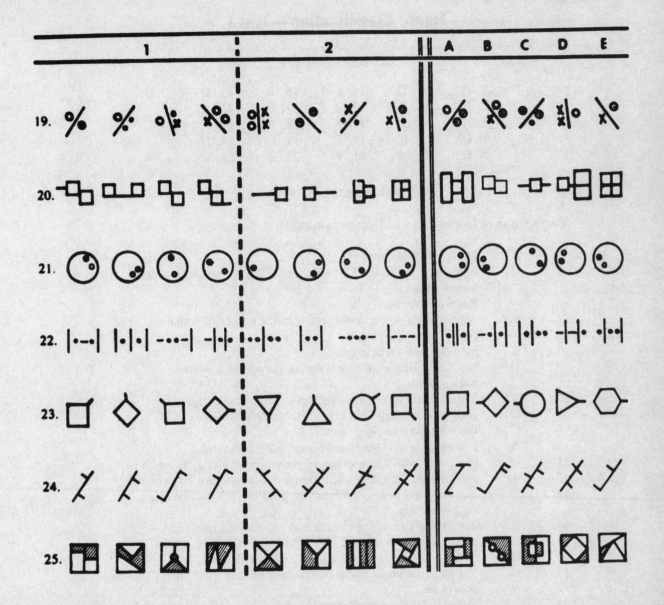

Figure Classification—Test I

Correct Answers

1. B	6. D	11. D	16. B	21. D
2. C	7. B	12. A	17. D	22. A
3. A	8. D	13. E	18. B	23. E
4. C	9. C	14. D	19. C	24. D
5. E	10. A	15. E	20. B	25. C

Explanatory Answers

Every figure in Group I, but no figure in Group II . . .

1. B has a point on top.
2. C contains a forward "S" (which may be on its side, but not a mirror image).
3. A has a dot above.
4. C consists of a single white figure in the center of a shaded figure.
5. E includes one vertical line.
6. D is a *single* figure (of any color or shape) on a white background.
7. B includes no right or obtuse angles (only acute angles).
8. D is divided equally between white and black areas.
9. C includes only straight lines.
10. A is a circle with "pie-shaped" sector(s) removed.
11. D is a circle with a line or curve running completely through it.
12. A is a triangle with one side extended, and one dot anywhere.
13. E is a rectangle with a different-colored circle attached to its rightmost side.
14. D consists of two white circles and one shaded circle.
15. E ends on a down stroke: ‾‾‾‾¬ (at the rightmost end).
16. B consists of two horizontal lines and one diagonal line.
17. D has an acute angle going *clockwise* from the long "hand" to the short one.
18. B includes four (and only four) vertical lines.
19. C has no two adjacent protrusions on the same side of the line.
20. B consists of two circles and two rectangles (only).
21. D has three horizontal lengths (between verticals).
22. A has the same number of dots on each side of the line.
23. E has more white boxes than shaded ones.
24. D has an odd number of lines.
25. C has the parts arranged so that all circles come to the left of everything else, all squares come to the left of triangles and dots, and all triangles precede dots.

Figure Classification—Test II

Correct Answers

1. C	6. A	11. D	16. C	21. E
2. D	7. C	12. A	17. C	22. B
3. B	8. D	13. E	18. A	23. B
4. B	9. E	14. E	19. E	24. E
5. D	10. E	15. B	20. B	25. C

Explanatory Answers

Every figure in Group I, but no figure in Group II . . .

1.	C	consists of three lines and three dots.
2.	D	consists of a single undivided shaded region touching any number of white regions.
3.	B	is a circle with three radii, two solid and one dotted such that two angles are formed totaling 180°.
4.	B	contains intersections in the middle of its leftmost and rightmost vertical lines.
5.	D	has three elements, the first and third being mirror images.
6.	A	has exactly one dot to the right of the line.
7.	C	has more white area than shaded area.
8.	D	has an even number of sides, with the dot outside the figure.
9.	E	is *a*symmetrical if a vertical line is drawn through the center.
10.	E	has one more vertical line than the number of horizontal lines.
11.	D	consists of one quadrilateral inscribed inside another quadrilateral.
12.	A	has an odd number of elements, with no dot above any line.
13.	E	consists of two congruent parts.
14.	E	has equal areas of white and shaded territory.
15.	B	has no vertical lines.
16.	C	is a hexagon with one dot inside and one dot outside.
17.	C	has an even number of dots.
18.	A	has ⌐ at its leftmost end.
19.	E	has two elements to the right of the line and one to the left of it (a dot in a circle together count as two elements).
20.	B	has two rectangles.
21.	E	has the black dot above the white dot.
22.	B	contains five parts, two of which are dots.
23.	B	is a quadrilateral with a line attached to it and extending horizontally, straight up, or diagonally up.
24.	E	is a line slanted like this / , with two perpendicular lines attached to it.
25.	C	consists of two shaded regions and two white ones.

SERIES REASONING TESTS

Numbers and letters may also be considered symbols when they are used to test your ability to think without words. This chapter provides a complete explanation of both number and letter series questions followed by practice tests with detailed explanations to each question.

NUMBER SERIES

In Number Series, each question consists of a group of numbers which follow some definite order. You must examine the numbers closely to determine the rule that governs the formation of the given series. Then you must use this rule to select the one of the lettered answer choices that will be the next number (or two numbers) in the series. The five sample questions that follow illustrate some of the variations possible with this question type.

1. 2 4 6 8 10 12
   - (A) 14
   - (B) 16
   - (C) 18
   - (D) 20
   - (E) none of these

In question 1, the rule is to add 2 to each number (2 + 2 = 4; 4 + 2 = 6; etc.) The next number in the series is 14 (12 + 2 = 14). Since 14 is lettered (A), (A) is your correct answer.

2. 7 8 6 7 5 6
   - (A) 2
   - (B) 3
   - (C) 4
   - (D) 5
   - (E) none of these

In question 2, the rule is to add 1 to the first number, subtract 2 from the next, add 1, subtract 2, and so on. The next number in the series is 4, letter (C) is your correct answer.

3. 20 20 21 21 22 22 23
   - (A) 23 23
   - (B) 23 24
   - (C) 19 19
   - (D) 22 23
   - (E) 21 22

In this series each number is repeated and then increased by 1. The rule is "repeat, add 1, repeat, add 1, etc." The series would be $20 +^0 20 +^1 21 +^0 21 +^1 22 +^0 22 +^1 23 +^0 23 +^1 24$. The correct answer is 23 and 24, which is alternative B.

4. 17 3 17 4 17 5 17
   - (A) 6 17
   - (B) 6 7
   - (C) 17 6
   - (D) 5 6
   - (E) 17 7

If you can't find a single rule for all the numbers in a series, see if there are really two series in the problem. This series is the number 17 separated by numbers increasing by 1, starting with 3. If the series were continued for two more numbers, it would read 17 3 17 4 17 5 17 6 17. The correct answer is 6 and 17, which is alternative A.

5. 1 2 4 5 7 8 10
   - (A) 11 12
   - (B) 12 14
   - (C) 10 13
   - (D) 12 13
   - (E) 11 13

The rule in this series is not easy to see until you actually set down how the numbers are changing: $1 +^1 2 +^2 4 +^1 5 +^2 7 +^1 8 +^2 10$. The numbers in this series are increasing first by 1 (that is plus 1) and then by 2 (that is plus 2). If the series were continued for two more numbers, it would read: 1 2 4 5 7 8 10 (plus 1) which is *11* (plus 2) which is *13*. Therefore the correct answer is 11 and 13, which is alternative E.

HINTS FOR ANSWERING
NUMBER SERIES QUESTIONS

● Do the ones that are easiest for you first. Then go back and work on the others. Enough time is allowed for you to do all the questions, providing you don't stay too long on the ones you have trouble answering.

● Sound out the series to yourself. You may hear the rule: 2 4 6 8 10 12 14 . . . What are the next two numbers?

● Look at the series carefully. You may see the rule: 9 2 9 4 9 6 9 . . . What are the next two numbers?

● If you can't hear it or see it, you may have to figure it out by writing down how the numbers are changing: 6 8 16 18 26 28 36 . . . What are the next two numbers? 6 + ^{2} 8 + ^{8} 16 + ^{2} 18 + ^{8} 26 + ^{2} 28 + ^{8} 36 . . . 36 + 2 = 38 + 8 = 46 or 38 46. You would mark the letter of the answer that goes with 38 46.

● If none of the answers given fit the rule you have figured out, try again. Try to figure out a rule that makes one of the five answers a correct one.

DON'T SPEND TOO MUCH TIME ON ANY ONE QUESTION.
SKIP IT AND COME BACK. A FRESH LOOK SOMETIMES HELPS.

LETTER SERIES

In Letter Series, each question consists of a number of letters which are arranged according to a definite pattern. You must discover what that pattern is and then use that knowledge to determine which of the five alternatives offered is the next letter in the series. The questions range in difficulty from simple alphabetical progressions to intricate combinations of progressions that alternate between forward and backward steps.

Since each question is based on the twenty-six letters of the alphabet, it is a good idea to keep a copy of the alphabet in front of you as you work. Additional-

ly, it is well worth your time to assign a number to each letter, jotting down the numbers from one to twenty-six directly under the letters to which they correspond. The seconds spent doing this may save you precious minutes as you work through the test, thus enabling you to solve extra problems and earn extra points.

There are three methods of attack for Letter Series questions. You may solve these problems by inspection, by diagraming or by numerical analysis. Each method has its advantages as you will see in the sample questions below.

INSPECTION

The first line of attack should always be inspection, for this is the quickest and easiest approach. Look at the letters. Are they progressing in normal or reverse alphabetical order? Are the letters consecutive or do they skip one or more letters between terms? Are certain letters repeated? Here are two simple series which you should be able to solve by inspection only.

1. c a d a e a f a g a
   (A) a   (D) b
   (B) g   (E) i
   (C) h
2. a b c c d e f f g h i
   (A) c
   (B) f

In the first sample question, the letters are progressing in consecutive alphabetical order with the letter /a/ inserted between each step. The next letter in this series must, therefore, be /h/ which is alternative C.

The second sample question is also a consecutive alphabetical progression, but here the third letter of each set is repeated. Thus we have: abcc deff ghii. Since only one /i/ is given in the original series, the next letter must be the second /i/ needed to complete the third set. Therefore, the answer, as determined by inspection only, is alternative E.

DIAGRAMING

When no pattern emerges by inspection alone, try diagraming the progression by linking the letters with pencil lines extending above the letters for each forward progression and below the letters for each backward progression. Sample questions 3 and 4 are good examples of how diagrams can help solve letter series problems.

3. a e c g e i g k i m k
   (A) i   (D) p
   (B) o   (E) n
   (C) m

4. m k o i q g s e u c
   (A) w   (D) b
   (B) x   (E) z
   (C) a

The letters in question 3 are obviously proceeding both forward and backward in the alphabet, and no progression is immediately apparent by inspection alone. A diagram of this question would look like this:

Now the pattern becomes obvious as a progression that goes ahead four letters and then back two letters. The missing letter is, therefore, four letters after /k/, which is the /o/ indicated by the dotted line in the diagram. Thus alternative B is correct.

As in the previous example, no pattern immediately emerges by inspection so a diagram is indicated.

The diagram makes it clear that this progression is orbiting around the letter /l/, first moving ahead one letter, then back one letter; next moving ahead three letters and back three letters; then moving ahead five letters and back five letters, etc. The dotted line indicates the next move in this series must be to /w/, which is eleven letters ahead of /l/. Therefore, alternative A is correct.

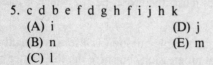

NUMERICAL ANALYSIS

If inspection and diagraming do not make the answer apparent, switch to a numerical analysis of the series. Assign each letter a numerical value according to its position in the alphabet. Once this is done, a pattern of pluses and minuses may emerge as in the following sample questions.

5. c d b e f d g h f i j h k
   (A) i   (D) j
   (B) n   (E) m
   (C) l

Assigning numerical values to each letter in this series we have:

```
c  d  b  e  f  d  g  h  f  i   j   h   k  or
3  4  2  5  6  4  7  8  6  9   10  8   11
 +1 -2 +3 +1 -2 +3 +1 -2 +3 +1  -2  +3
```

Now it is obvious that the series is progressing by the formula +1 -2 +3. According to this pattern, the next letter must be 11 + 1 or 12 which corresponds to the letter /l/ as in alternative C.

6. a c g f c e i h e g k j
   (A) l   (D) g
   (B) n   (E) m
   (C) i

Once again assigning numerical values to each letter this series becomes:

```
a   c   g   f   c   e   i   h   e   g   k   j
1   3   7   6   3   5   9   8   5   7   11  10
 +2  +4  -1  -3  +2  +4  -1  -3  +2  +4  -1
```

Now you can see that the pattern is +2 +4 -1 -3 which repeats throughout the series. Therefore, the next term must be 10 -3 which is 7, corresponding to the letter /g/ or alternative D.

Now go on to try the practice test that follows. Check the explanations for any questions you answer incorrectly.

TEST I. NUMBER SERIES

DIRECTIONS: Each question consists of a series of numbers which follow some definite order. Study each series to determine what the order is. Then look at the answer choices. Select the one answer that will complete the set in accordance with the pattern established.

1. 8 12 17 24 28 33
   (A) 36   (D) 39
   (B) 37   (E) 40
   (C) 38

2. 3 12 6 24 12 48
   (A) 24   (D) 40
   (B) 32   (E) 64
   (C) 36

3. 7 11 16 22 26 31
   (A) 32   (D) 37
   (B) 34   (E) 39
   (C) 36

4. 24 12 36 24 48 36
   (A) 40   (D) 58
   (B) 50   (E) 60
   (C) 52

5. 15 13 11 14 17 15
   (A) 11   (D) 14
   (B) 12   (E) 16
   (C) 13

6. 8 7 10 5 4 7
   (A) 6   (D) 2
   (B) 4   (E) 1
   (C) 3

7. 15 11 7 14 10 6
   (A) 4   (D) 10
   (B) 6   (E) 12
   (C) 8

8. 7 4 12 9 27 24
   (A) 11   (D) 72
   (B) 36   (E) 96
   (C) 48

9. 5 3 9 7 21 19
   (A) 9   (D) 64
   (B) 36   (E) 72
   (C) 57

10. 11 8 16 17 14 28
    (A) 20   (D) 38
    (B) 29   (E) 40
    (C) 32

11. 64 32 16 8 4 2
    (A) 1/8   (D) 1
    (B) 1/4   (E) 0
    (C) 1/2

12. 48 24 20 10 6 3
    (A) 2   (D) —1
    (B) 1   (E) —2
    (C) 0

13. 77 76 74 71 67 62
    (A) 60   (D) 57
    (B) 59   (E) 56
    (C) 58

14. 4 2 8 4 12 6
    (A) 8   (D) 11
    (B) 9   (E) 12
    (C) 10

15. 1 6 36 3 18 108
    (A) 7   (D) 10
    (B) 8   (E) 11
    (C) 9

16. 2 6 12 36 72 216
    (A) 288   (D) 476
    (B) 376   (E) 648
    (C) 432

17. .05 .1 .3 1.2 6 36
    (A) 48   (D) 216
    (B) 72   (E) 252
    (C) 164

18. 7 6 1/2 6 1/4 5 3/4 5 1/2 5
    (A) 4 3/4   (D) 4
    (B) 4 1/2   (E) 3 3/4
    (C) 4 1/4

19. 1 2 4 8 16 32
    (A) 48   (D) 80
    (B) 56   (E) 96
    (C) 64

20. 4 9 16 25 36 49
    (A) 51   (D) 60
    (B) 53   (E) 64
    (C) 54

21. 17 19 22 26 31 37
    (A) 40   (D) 43
    (B) 41   (E) 44
    (C) 42

22. 1/16 1/4 1/2 2 4 16
    (A) 24   (D) 64
    (B) 32   (E) 80
    (C) 48

23. 1 2 3 2 3 4
    (A) 3
    (B) 4
    (C) 5
    (D) 6
    (E) 7

24. 3 9 14 18 21 23
    (A) 24
    (B) 25
    (C) 26
    (D) 27
    (E) 28

25. 90 45 50 25 30 15
    (A) 16
    (B) 18
    (C) 20
    (D) 10
    (E) 5

TEST II. LETTER SERIES

DIRECTIONS: Each question consists of a series of letters or numbers (or both) which follow some definite order. Study each series to determine what the order is. Then look at the answer choices. Select the one answer that will complete the set in accordance with the pattern established.

Suggestions: In solving alphabetic series, it is helpful to write out the alphabet and keep it in front of you as you work. This makes it easier to spot the key to a letter series.

A	B	C	D	E	F	G	H	I	J	K	L	M	N	O	P	Q	R	S	T	U	V	W	X	Y	Z
1	2	3	4	5	6	7	8	9	10	11	12	13	14	15	16	17	18	19	20	21	22	23	24	25	26

1. a c e b d f c e g d f h e
   - (A) g
   - (B) h
   - (C) i
   - (D) d
   - (E) f

2. f g i h j k m l n o q p r s
   - (A) t
   - (B) u
   - (C) v
   - (D) r
   - (E) p

3. m p n o p p q r p s t u
   - (A) p
   - (B) s
   - (C) v
   - (D) w
   - (E) x

4. t t s r r q p p o n n m l l
   - (A) j
   - (B) l
   - (C) n
   - (D) m
   - (E) k

5. m n a z l o b y k p c x j
   - (A) w
   - (B) j
   - (C) d
   - (D) q
   - (E) i

6. g g h h h i i i i j j k k
   - (A) l
   - (B) j
   - (C) k
   - (D) m
   - (E) n

7. c d r s h i l m f g x y p
   - (A) a
   - (B) r
   - (C) q
   - (D) z
   - (E) w

8. b a z d c y f e x h g
   - (A) k
   - (B) w
   - (C) l
   - (D) p
   - (E) r

9. l m n o p q q r s t t t u v w w
   - (A) w
   - (B) v
   - (C) p
   - (D) x
   - (E) l

10. p a c d p e f h p i j m p o n
    - (A) o
    - (B) p
    - (C) q
    - (D) r
    - (E) s

11. b e g g b c o d d c d a l l d
    - (A) d
    - (B) e
    - (C) f
    - (D) i
    - (E) u

12. f o r f o p r f o p q r f o p q ɪ
    - (A) f
    - (B) o
    - (C) u
    - (D) r
    - (E) t

13. f h j l n p r t v x
    - (A) w
    - (B) s
    - (C) y
    - (D) z
    - (E) u

14. a a b b d d g g o o p p
    - (A) h
    - (B) q
    - (C) t
    - (D) u
    - (E) i

15. y j j b j j w j j d j j u j j
    - (A) f
    - (B) j
    - (C) e
    - (D) c
    - (E) x

Number Series—Test I

Correct Answers

1. E	6. D	11. D	16. C	21. E
2. A	7. E	12. D	17. E	22. B
3. D	8. D	13. E	18. A	23. A
4. E	9. C	14. E	19. C	24. A
5. C	10. B	15. C	20. E	25. C

Explanatory Answers

1.   E   To obtain the terms in the series, take the first term—then add 4, add 5, add 7; repeat the cycle.

2.   A   Multiply by 4, divide by 2; repeat the cycle.

3.   D   Add 4, add 5, add 6; repeat the cycle.

4.   E   Subtract 12, add 24; repeat the cycle.

5.   C   Subtract 2, subtract 2, add 3, add 3; repeat the cycle.

6.   D   Subtract 1, add 3, subtract 5; repeat the cycle.

7.   E   Subtract 4, subtract 4, multiply by 2; repeat the cycle.

8.   D   Subtract 3, multiply by 3; repeat the cycle.

9.   C   Subtract 2, multiply by 3; repeat the cycle.

10.   B   Subtract 3, multiply by 2, add 1; repeat the cycle.

11.   D   Divide by 2; repeat.

12.   D   Divide by 2, subtract 4; repeat the cycle.

13.   E   Subtract 1, subtract 2, subtract 3, subtract 4 . . . (subtract one more each time).

14.   E   Divide by 2, multiply by 4; divide by 2, multiply by 3; divide by 2, multiply by 2; etc. (repeat, subtracting one from the underlined number each time).

15.   C   Multiply by 6, multiply by 6, divide by 12; repeat the cycle.

16.   C   Multiply by 3, multiply by 2; repeat the cycle.

17.   E   Multiply by 2; multiply by 3; multiply by 4; etc. (repeat, adding one to the underlined number each time).

18.   A   Subtract 1/2, subtract 1/4; repeat cycle.

19.   C   Multiply by 2; repeat.

20.   E   Add 5; add 7; add 9; etc. (repeat, adding two to the underlined number each time).

21.   E   Add 2; add 3; add 4; etc. (repeat, adding one to the underlined number each time).

22.   B   Multiply by 4, multiply by 2; repeat the cycle.

23.   A   Add 1, add 1, subtract 1; repeat the cycle.

24.   A   Add 6; add 5; add 4; etc. (repeat, subtracting one from the underlined number each time).

25.   C   Divide by 2, add 5; repeat the cycle.

TEST II. LETTER SERIES

CORRECT ANSWERS

1. A	4. E	7. C	10. D	13. D
2. B	5. D	8. B	11. B	14. B
3. C	6. C	9. A	12. D	15. A

EXPLANATORY ANSWERS

1. **(A)** The series consists of three-letter sets which follow in alphabetical order but skip one letter between each term. Each new set begins with the letter immediately following the first letter of the previous set. Thus, the letter /e/ is the start of the fifth three-letter set and must be followed by /g/.

2. **(B)** This series consists of four-letter sets in which the first two letters of each set are in alphabetical order and the second two letters are in reverse alphabetical order. Thus /fg/ is followed by /ih/; /jk/ by /ml/, /no/ by /qp/; and /rs/ by /ut/.

3. **(C)** The letter /p/ is followed by alphabetical series which increase by one letter each time, starting with the single letter /m/. Thus the first /p/ is followed by the two letters /no/; the second /p/ is followed by the three letters /pqr/; and the third /p/ must be followed by four letters /stuv/.

4. **(E)** In this series the letters are in reverse alphabetical order following the pattern of one letter appearing twice and the next letter appearing once throughout the series. Thus the series is: tt s rr q pp o nn m ll. The next letter must be the single letter /k/, which is one letter back from /l/.

5. **(D)** A diagram of this series reveals that the pattern alternates between letters radiating out from the middle of the alphabet and letters moving in from the two ends of the alphabet as illustrated below:

6. **(C)** In this series, the letters are in alphabetical order with the first letter appearing twice, the second letter appearing three times and the third letter appearing four times. Then the series reverts back so that the fourth letter appears twice; therefore, the fifth letter should appear three times as follows: gg hhh iiii jj kkk.

7. **(C)** The pattern established by this series is a random assortment of any two letters in alphabetical sequence. Therefore, the next letter must be the letter that immediately follows /p/, which, of course, is /q/. This pattern is easily seen when the letters are regrouped as follows: cd rs hi lm fg xy pq.

8. **(B)** This pattern is formed by reversing the order of two letters from the beginning of the alphabet and following them with one letter from the end of the alphabet, then repeating the pattern with the next available letters from each end of the alphabet. Thus: baz dcy fex hgw.

9. **(A)** This pattern consists of three consecutive letters of the alphabet. However, each third letter multiplies itself, increasing by one in each set. Thus we have: lm/n/ op/qq/ rs/ttt/ uv/wwww/.

10. **(D)** In this pattern each /p/ is followed by a vowel and then by two letters with a constantly increasing gap between the letters. Thus, /pa/ is followed by /cd/ (c + 1 = d); /pe/ is followed by /fh/ (f + 2 = h); /pi/ is followed by /jm/ (j + 3 = m); and /po/ must be followed by /nr/ (n + 4 = r).

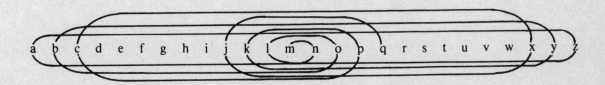

The next letter must be the /q/ indicated by the line above.

11. **(B)** This series consists of five-letter sets that begin and end with the same letter. Each five-letter set begins and ends with letters in consecutive alphabetical order. Thus, the first set begins and ends with /b/ (beggb), the second set begins and ends with /c/ (coddc) and the third set begins and ends with /d/ (dalld). The next set must, therefore, begin and end with /e/.

12. **(D)** This pattern consists of an increasing alphabetic progression bounded by the letter /f/ and /r/. Thus, the first set consists of /f/ followed by the single letter /o/ and then /r/. The second set is /f/ followed by the two letters /op/ and then /r/. The third set is /f/ followed by the three letters /opq/ and then /r/. The fourth set must, therefore, be /f/ followed by the four letters /opqr/, and the next letter is the /r/ needed to complete the set.

13. **(D)** The pattern here consists of skipping every other letter:
f (g) h (i) j (k) l (m) n (o) p (q) r (s) - t (u) v (w) x. The next letter must, therefore, be /z/.

14. **(B)** The letters /a b d g o p/ have one thing in common — each contains a closed loop. The only other letter with a closed loop is /q/, which must, therefore be the next letter in this series.

15. **(A)** This series begins at the end of the alphabet then goes to the beginning of the alphabet choosing only every other letter. A double /j/ is inserted between each term.

Part Three
Mechanical Aptitude
and Achievement

PART THREE

MECHANICAL APTITUDE AND ACHIEVEMENT
Answer Sheets

To simulate actual examination conditions, mark your answers to each question in this chapter on these answer sheets. Make one clear, black mark for each answer. If you decide to change an answer, erase your error completely. On machine-scored examinations, additional or stray marks on your answer sheet may be counted as mistakes. After you have taken all the tests, compare your answers to the Correct Answers at the end of the chapter to see where you stand.

TOOL KNOWLEDGE

Tool Recognition—Test I

1. _____   4. _____   7. _____   10. _____

2. _____   5. _____   8. _____   11. _____

3. _____   6. _____   9. _____

Tool Recognition—Test II

1. _____   9. _____   17. _____   25. _____

2. _____   10. _____   18. _____   26. _____

3. _____   11. _____   19. _____   27. _____

4. _____   12. _____   20. _____   28. _____

5. _____   13. _____   21. _____   29. _____

6. _____   14. _____   22. _____   30. _____

7. _____   15. _____   23. _____   31. _____

8. _____   16. _____   24. _____

Tool Recognition—Test III

1. _____	7. _____	13. _____	19. _____
2. _____	8. _____	14. _____	20. _____
3. _____	9. _____	15. _____	21. _____
4. _____	10. _____	16. _____	22. _____
5. _____	11. _____	17. _____	23. _____
6. _____	12. _____	18. _____	24. _____

Tool Recognition—Test IV

1. _____	4. _____	7. _____	10. _____
2. _____	5. _____	8. _____	11. _____
3. _____	6. _____	9. _____	

Tool Recognition—Test V

1. _____	4. _____	7. _____	10. _____
2. _____	5. _____	8. _____	11. _____
3. _____	6. _____	9. _____	12. _____

Tool Analogy—Test VI

1 Ⓐ Ⓑ Ⓒ Ⓓ	3 Ⓐ Ⓑ Ⓒ Ⓓ	5 Ⓐ Ⓑ Ⓒ Ⓓ	7 Ⓐ Ⓑ Ⓒ Ⓓ
2 Ⓐ Ⓑ Ⓒ Ⓓ	4 Ⓐ Ⓑ Ⓒ Ⓓ	6 Ⓐ Ⓑ Ⓒ Ⓓ	8 Ⓐ Ⓑ Ⓒ Ⓓ

Tool Analogy Test—VII

1 Ⓐ Ⓑ Ⓒ Ⓓ	3 Ⓐ Ⓑ Ⓒ Ⓓ	5 Ⓐ Ⓑ Ⓒ Ⓓ	7 Ⓐ Ⓑ Ⓒ Ⓓ
2 Ⓐ Ⓑ Ⓒ Ⓓ	4 Ⓐ Ⓑ Ⓒ Ⓓ	6 Ⓐ Ⓑ Ⓒ Ⓓ	8 Ⓐ Ⓑ Ⓒ Ⓓ

Tool Analogy—Test VIII

1 (A)(B)(C)(D)   3 (A)(B)(C)(D)   5 (A)(B)(C)(D)   7 (A)(B)(C)(D)

2 (A)(B)(C)(D)   4 (A)(B)(C)(D)   6 (A)(B)(C)(D)   8 (A)(B)(C)(D)

MECHANICAL INSIGHT

Mechanical Insight—Test I

1 (A)(B)(C)(D)(E)   7 (A)(B)(C)(D)(E)   13 (A)(B)(C)(D)(E)

2 (A)(B)(C)(D)(E)   8 (A)(B)(C)(D)(E)   14 (A)(B)(C)(D)(E)

3 (A)(B)(C)(D)(E)   9 (A)(B)(C)(D)(E)   15 (A)(B)(C)(D)(E)

4 (A)(B)(C)(D)(E)   10 (A)(B)(C)(D)(E)   16 (A)(B)(C)(D)(E)

5 (A)(B)(C)(D)(E)   11 (A)(B)(C)(D)(E)   17 (A)(B)(C)(D)(E)

6 (A)(B)(C)(D)(E)   12 (A)(B)(C)(D)(E)   18 (A)(B)(C)(D)(E)

Mechanical Insight—Test II

1 (A)(B)(C)(D)   5 (A)(B)(C)(D)   9 (A)(B)(C)(D)   13 (A)(B)(C)(D)

2 (A)(B)(C)(D)   6 (A)(B)(C)(D)   10 (A)(B)(C)(D)   14 (A)(B)(C)(D)

3 (A)(B)(C)(D)   7 (A)(B)(C)(D)   11 (A)(B)(C)(D)   15 (A)(B)(C)(D)

4 (A)(B)(C)(D)   8 (A)(B)(C)(D)   12 (A)(B)(C)(D)   16 (A)(B)(C)(D)

Mechanical Insight—Test III

1 (A)(B)(C)(D)   5 (A)(B)(C)(D)   9 (A)(B)(C)(D)   13 (A)(B)(C)(D)

2 (A)(B)(C)(D)   6 (A)(B)(C)(D)   10 (A)(B)(C)(D)   14 (A)(B)(C)(D)

3 (A)(B)(C)(D)   7 (A)(B)(C)(D)   11 (A)(B)(C)(D)   15 (A)(B)(C)(D)

4 (A)(B)(C)(D)   8 (A)(B)(C)(D)   12 (A)(B)(C)(D)   16 (A)(B)(C)(D)

Mechanical Insight—Test IV

1 Ⓐ Ⓑ Ⓒ Ⓓ Ⓔ   5 Ⓐ Ⓑ Ⓒ Ⓓ Ⓔ   9 Ⓐ Ⓑ Ⓒ Ⓓ Ⓔ   13 Ⓐ Ⓑ Ⓒ Ⓓ Ⓔ   17 Ⓐ Ⓑ Ⓒ Ⓓ Ⓔ

2 Ⓐ Ⓑ Ⓒ Ⓓ Ⓔ   6 Ⓐ Ⓑ Ⓒ Ⓓ Ⓔ   10 Ⓐ Ⓑ Ⓒ Ⓓ Ⓔ   14 Ⓐ Ⓑ Ⓒ Ⓓ Ⓔ   18 Ⓐ Ⓑ Ⓒ Ⓓ Ⓔ

3 Ⓐ Ⓑ Ⓒ Ⓓ Ⓔ   7 Ⓐ Ⓑ Ⓒ Ⓓ Ⓔ   11 Ⓐ Ⓑ Ⓒ Ⓓ Ⓔ   15 Ⓐ Ⓑ Ⓒ Ⓓ Ⓔ   19 Ⓐ Ⓑ Ⓒ Ⓓ Ⓔ

4 Ⓐ Ⓑ Ⓒ Ⓓ Ⓔ   8 Ⓐ Ⓑ Ⓒ Ⓓ Ⓔ   12 Ⓐ Ⓑ Ⓒ Ⓓ Ⓔ   16 Ⓐ Ⓑ Ⓒ Ⓓ Ⓔ

Mechanical Insight—Test V

1 Ⓐ Ⓑ Ⓒ Ⓓ   6 Ⓐ Ⓑ Ⓒ Ⓓ   11 Ⓐ Ⓑ Ⓒ Ⓓ   16 Ⓐ Ⓑ Ⓒ Ⓓ

2 Ⓐ Ⓑ Ⓒ Ⓓ   7 Ⓐ Ⓑ Ⓒ Ⓓ   12 Ⓐ Ⓑ Ⓒ Ⓓ   17 Ⓐ Ⓑ Ⓒ Ⓓ

3 Ⓐ Ⓑ Ⓒ Ⓓ   8 Ⓐ Ⓑ Ⓒ Ⓓ   13 Ⓐ Ⓑ Ⓒ Ⓓ   18 Ⓐ Ⓑ Ⓒ Ⓓ

4 Ⓐ Ⓑ Ⓒ Ⓓ   9 Ⓐ Ⓑ Ⓒ Ⓓ   14 Ⓐ Ⓑ Ⓒ Ⓓ   19 Ⓐ Ⓑ Ⓒ Ⓓ

5 Ⓐ Ⓑ Ⓒ Ⓓ   10 Ⓐ Ⓑ Ⓒ Ⓓ   15 Ⓐ Ⓑ Ⓒ Ⓓ   20 Ⓐ Ⓑ Ⓒ Ⓓ

MECHANICAL KNOWLEDGE

Shop Practices—Test I

1 Ⓐ Ⓑ Ⓒ Ⓓ   6 Ⓐ Ⓑ Ⓒ Ⓓ   11 Ⓐ Ⓑ Ⓒ Ⓓ   16 Ⓐ Ⓑ Ⓒ Ⓓ

2 Ⓐ Ⓑ Ⓒ Ⓓ   7 Ⓐ Ⓑ Ⓒ Ⓓ   12 Ⓐ Ⓑ Ⓒ Ⓓ   17 Ⓐ Ⓑ Ⓒ Ⓓ

3 Ⓐ Ⓑ Ⓒ Ⓓ   8 Ⓐ Ⓑ Ⓒ Ⓓ   13 Ⓐ Ⓑ Ⓒ Ⓓ   18 Ⓐ Ⓑ Ⓒ Ⓓ

4 Ⓐ Ⓑ Ⓒ Ⓓ   9 Ⓐ Ⓑ Ⓒ Ⓓ   14 Ⓐ Ⓑ Ⓒ Ⓓ   19 Ⓐ Ⓑ Ⓒ Ⓓ

5 Ⓐ Ⓑ Ⓒ Ⓓ   10 Ⓐ Ⓑ Ⓒ Ⓓ   15 Ⓐ Ⓑ Ⓒ Ⓓ   20 Ⓐ Ⓑ Ⓒ Ⓓ

Electronics Information—Test II

1 Ⓐ Ⓑ Ⓒ Ⓓ   7 Ⓐ Ⓑ Ⓒ Ⓓ   13 Ⓐ Ⓑ Ⓒ Ⓓ   19 Ⓐ Ⓑ Ⓒ Ⓓ   25 Ⓐ Ⓑ Ⓒ Ⓓ

2 Ⓐ Ⓑ Ⓒ Ⓓ   8 Ⓐ Ⓑ Ⓒ Ⓓ   14 Ⓐ Ⓑ Ⓒ Ⓓ   20 Ⓐ Ⓑ Ⓒ Ⓓ   26 Ⓐ Ⓑ Ⓒ Ⓓ

3 Ⓐ Ⓑ Ⓒ Ⓓ   9 Ⓐ Ⓑ Ⓒ Ⓓ   15 Ⓐ Ⓑ Ⓒ Ⓓ   21 Ⓐ Ⓑ Ⓒ Ⓓ   27 Ⓐ Ⓑ Ⓒ Ⓓ

4 Ⓐ Ⓑ Ⓒ Ⓓ   10 Ⓐ Ⓑ Ⓒ Ⓓ   16 Ⓐ Ⓑ Ⓒ Ⓓ   22 Ⓐ Ⓑ Ⓒ Ⓓ   28 Ⓐ Ⓑ Ⓒ Ⓓ

5 Ⓐ Ⓑ Ⓒ Ⓓ   11 Ⓐ Ⓑ Ⓒ Ⓓ   17 Ⓐ Ⓑ Ⓒ Ⓓ   23 Ⓐ Ⓑ Ⓒ Ⓓ   29 Ⓐ Ⓑ Ⓒ Ⓓ

6 Ⓐ Ⓑ Ⓒ Ⓓ   12 Ⓐ Ⓑ Ⓒ Ⓓ   18 Ⓐ Ⓑ Ⓒ Ⓓ   24 Ⓐ Ⓑ Ⓒ Ⓓ   30 Ⓐ Ⓑ Ⓒ Ⓓ

Automotive Information—Test III

1 Ⓐ Ⓑ Ⓒ Ⓓ   5 Ⓐ Ⓑ Ⓒ Ⓓ   9 Ⓐ Ⓑ Ⓒ Ⓓ   13 Ⓐ Ⓑ Ⓒ Ⓓ   17 Ⓐ Ⓑ Ⓒ Ⓓ

2 Ⓐ Ⓑ Ⓒ Ⓓ   6 Ⓐ Ⓑ Ⓒ Ⓓ   10 Ⓐ Ⓑ Ⓒ Ⓓ   14 Ⓐ Ⓑ Ⓒ Ⓓ   18 Ⓐ Ⓑ Ⓒ Ⓓ

3 Ⓐ Ⓑ Ⓒ Ⓓ   7 Ⓐ Ⓑ Ⓒ Ⓓ   11 Ⓐ Ⓑ Ⓒ Ⓓ   15 Ⓐ Ⓑ Ⓒ Ⓓ   19 Ⓐ Ⓑ Ⓒ Ⓓ

4 Ⓐ Ⓑ Ⓒ Ⓓ   8 Ⓐ Ⓑ Ⓒ Ⓓ   12 Ⓐ Ⓑ Ⓒ Ⓓ   16 Ⓐ Ⓑ Ⓒ Ⓓ   20 Ⓐ Ⓑ Ⓒ Ⓓ

Maintenance Work—Test IV

1 Ⓐ Ⓑ Ⓒ Ⓓ   12 Ⓐ Ⓑ Ⓒ Ⓓ   23 Ⓐ Ⓑ Ⓒ Ⓓ   34 Ⓐ Ⓑ Ⓒ Ⓓ   45 Ⓐ Ⓑ Ⓒ Ⓓ

2 Ⓐ Ⓑ Ⓒ Ⓓ   13 Ⓐ Ⓑ Ⓒ Ⓓ   24 Ⓐ Ⓑ Ⓒ Ⓓ   35 Ⓐ Ⓑ Ⓒ Ⓓ   46 Ⓐ Ⓑ Ⓒ Ⓓ

3 Ⓐ Ⓑ Ⓒ Ⓓ   14 Ⓐ Ⓑ Ⓒ Ⓓ   25 Ⓐ Ⓑ Ⓒ Ⓓ   36 Ⓐ Ⓑ Ⓒ Ⓓ   47 Ⓐ Ⓑ Ⓒ Ⓓ

4 Ⓐ Ⓑ Ⓒ Ⓓ   15 Ⓐ Ⓑ Ⓒ Ⓓ   26 Ⓐ Ⓑ Ⓒ Ⓓ   37 Ⓐ Ⓑ Ⓒ Ⓓ   48 Ⓐ Ⓑ Ⓒ Ⓓ

5 Ⓐ Ⓑ Ⓒ Ⓓ   16 Ⓐ Ⓑ Ⓒ Ⓓ   27 Ⓐ Ⓑ Ⓒ Ⓓ   38 Ⓐ Ⓑ Ⓒ Ⓓ   49 Ⓐ Ⓑ Ⓒ Ⓓ

6 Ⓐ Ⓑ Ⓒ Ⓓ   17 Ⓐ Ⓑ Ⓒ Ⓓ   28 Ⓐ Ⓑ Ⓒ Ⓓ   39 Ⓐ Ⓑ Ⓒ Ⓓ   50 Ⓐ Ⓑ Ⓒ Ⓓ

7 Ⓐ Ⓑ Ⓒ Ⓓ   18 Ⓐ Ⓑ Ⓒ Ⓓ   29 Ⓐ Ⓑ Ⓒ Ⓓ   40 Ⓐ Ⓑ Ⓒ Ⓓ   51 Ⓐ Ⓑ Ⓒ Ⓓ

8 Ⓐ Ⓑ Ⓒ Ⓓ   19 Ⓐ Ⓑ Ⓒ Ⓓ   30 Ⓐ Ⓑ Ⓒ Ⓓ   41 Ⓐ Ⓑ Ⓒ Ⓓ   52 Ⓐ Ⓑ Ⓒ Ⓓ

9 Ⓐ Ⓑ Ⓒ Ⓓ   20 Ⓐ Ⓑ Ⓒ Ⓓ   31 Ⓐ Ⓑ Ⓒ Ⓓ   42 Ⓐ Ⓑ Ⓒ Ⓓ   53 Ⓐ Ⓑ Ⓒ Ⓓ

10 Ⓐ Ⓑ Ⓒ Ⓓ   21 Ⓐ Ⓑ Ⓒ Ⓓ   32 Ⓐ Ⓑ Ⓒ Ⓓ   43 Ⓐ Ⓑ Ⓒ Ⓓ   54 Ⓐ Ⓑ Ⓒ Ⓓ

11 Ⓐ Ⓑ Ⓒ Ⓓ   22 Ⓐ Ⓑ Ⓒ Ⓓ   33 Ⓐ Ⓑ Ⓒ Ⓓ   44 Ⓐ Ⓑ Ⓒ Ⓓ   55 Ⓐ Ⓑ Ⓒ Ⓓ

SHOP ARITHMETIC

Shop Arithmetic—Test I

1 Ⓐ Ⓑ Ⓒ Ⓓ   10 Ⓐ Ⓑ Ⓒ Ⓓ   19 Ⓐ Ⓑ Ⓒ Ⓓ   28 Ⓐ Ⓑ Ⓒ Ⓓ

2 Ⓐ Ⓑ Ⓒ Ⓓ   11 Ⓐ Ⓑ Ⓒ Ⓓ   20 Ⓐ Ⓑ Ⓒ Ⓓ   29 Ⓐ Ⓑ Ⓒ Ⓓ

3 Ⓐ Ⓑ Ⓒ Ⓓ   12 Ⓐ Ⓑ Ⓒ Ⓓ   21 Ⓐ Ⓑ Ⓒ Ⓓ   30 Ⓐ Ⓑ Ⓒ Ⓓ

4 Ⓐ Ⓑ Ⓒ Ⓓ   13 Ⓐ Ⓑ Ⓒ Ⓓ   22 Ⓐ Ⓑ Ⓒ Ⓓ   31 Ⓐ Ⓑ Ⓒ Ⓓ

5 Ⓐ Ⓑ Ⓒ Ⓓ   14 Ⓐ Ⓑ Ⓒ Ⓓ   23 Ⓐ Ⓑ Ⓒ Ⓓ   32 Ⓐ Ⓑ Ⓒ Ⓓ

6 Ⓐ Ⓑ Ⓒ Ⓓ   15 Ⓐ Ⓑ Ⓒ Ⓓ   24 Ⓐ Ⓑ Ⓒ Ⓓ   33 Ⓐ Ⓑ Ⓒ Ⓓ

7 Ⓐ Ⓑ Ⓒ Ⓓ   16 Ⓐ Ⓑ Ⓒ Ⓓ   25 Ⓐ Ⓑ Ⓒ Ⓓ   34 Ⓐ Ⓑ Ⓒ Ⓓ

8 Ⓐ Ⓑ Ⓒ Ⓓ   17 Ⓐ Ⓑ Ⓒ Ⓓ   26 Ⓐ Ⓑ Ⓒ Ⓓ

9 Ⓐ Ⓑ Ⓒ Ⓓ   18 Ⓐ Ⓑ Ⓒ Ⓓ   27 Ⓐ Ⓑ Ⓒ Ⓓ

Shop Arithmetic—Test II

1 Ⓐ Ⓑ Ⓒ Ⓓ   8 Ⓐ Ⓑ Ⓒ Ⓓ   15 Ⓐ Ⓑ Ⓒ Ⓓ

2 Ⓐ Ⓑ Ⓒ Ⓓ   9 Ⓐ Ⓑ Ⓒ Ⓓ   16 Ⓐ Ⓑ Ⓒ Ⓓ

3 Ⓐ Ⓑ Ⓒ Ⓓ   10 Ⓐ Ⓑ Ⓒ Ⓓ   17 Ⓐ Ⓑ Ⓒ Ⓓ

4 Ⓐ Ⓑ Ⓒ Ⓓ   11 Ⓐ Ⓑ Ⓒ Ⓓ   18 Ⓐ Ⓑ Ⓒ Ⓓ

5 Ⓐ Ⓑ Ⓒ Ⓓ   12 Ⓐ Ⓑ Ⓒ Ⓓ   19 Ⓐ Ⓑ Ⓒ Ⓓ

6 Ⓐ Ⓑ Ⓒ Ⓓ   13 Ⓐ Ⓑ Ⓒ Ⓓ   20 Ⓐ Ⓑ Ⓒ Ⓓ

7 Ⓐ Ⓑ Ⓒ Ⓓ   14 Ⓐ Ⓑ Ⓒ Ⓓ   21 Ⓐ Ⓑ Ⓒ Ⓓ

Shop Arithmetic—Test III

1 Ⓐ Ⓑ Ⓒ Ⓓ   5 Ⓐ Ⓑ Ⓒ Ⓓ   9 Ⓐ Ⓑ Ⓒ Ⓓ   13 Ⓐ Ⓑ Ⓒ Ⓓ   17 Ⓐ Ⓑ Ⓒ Ⓓ

2 Ⓐ Ⓑ Ⓒ Ⓓ   6 Ⓐ Ⓑ Ⓒ Ⓓ   10 Ⓐ Ⓑ Ⓒ Ⓓ   14 Ⓐ Ⓑ Ⓒ Ⓓ   18 Ⓐ Ⓑ Ⓒ Ⓓ

3 Ⓐ Ⓑ Ⓒ Ⓓ   7 Ⓐ Ⓑ Ⓒ Ⓓ   11 Ⓐ Ⓑ Ⓒ Ⓓ   15 Ⓐ Ⓑ Ⓒ Ⓓ   19 Ⓐ Ⓑ Ⓒ Ⓓ

4 Ⓐ Ⓑ Ⓒ Ⓓ   8 Ⓐ Ⓑ Ⓒ Ⓓ   12 Ⓐ Ⓑ Ⓒ Ⓓ   16 Ⓐ Ⓑ Ⓒ Ⓓ   20 Ⓐ Ⓑ Ⓒ Ⓓ

Shop Arithmetic—Test IV

1 Ⓐ Ⓑ Ⓒ Ⓓ   5 Ⓐ Ⓑ Ⓒ Ⓓ   9 Ⓐ Ⓑ Ⓒ Ⓓ   13 Ⓐ Ⓑ Ⓒ Ⓓ   17 Ⓐ Ⓑ Ⓒ Ⓓ

2 Ⓐ Ⓑ Ⓒ Ⓓ   6 Ⓐ Ⓑ Ⓒ Ⓓ   10 Ⓐ Ⓑ Ⓒ Ⓓ   14 Ⓐ Ⓑ Ⓒ Ⓓ   18 Ⓐ Ⓑ Ⓒ Ⓓ

3 Ⓐ Ⓑ Ⓒ Ⓓ   7 Ⓐ Ⓑ Ⓒ Ⓓ   11 Ⓐ Ⓑ Ⓒ Ⓓ   15 Ⓐ Ⓑ Ⓒ Ⓓ   19 Ⓐ Ⓑ Ⓒ Ⓓ

4 Ⓐ Ⓑ Ⓒ Ⓓ   8 Ⓐ Ⓑ Ⓒ Ⓓ   12 Ⓐ Ⓑ Ⓒ Ⓓ   16 Ⓐ Ⓑ Ⓒ Ⓓ   20 Ⓐ Ⓑ Ⓒ Ⓓ

TOOL KNOWLEDGE TESTS

Since every craftsman should be familiar with the tools of his trade, and since familiarity with tools is an important measure of interest and motivation, tool knowledge questions have become an important feature of many mechanical ability, aptitude and comprehension tests. In order to familiarize you with this type of question, we have assembled a variety of tool knowledge questions from many different examinations. The more practice questions you answer, the better prepared you will be for your test.

There are two main types of questions testing your knowledge of tools — Tool Recognition questions and Tool Analogy questions.

In Tool Recognition Tests you are given a list of jobs to be done and an illustration showing as assortment of tools. The problem is to match the tool with the job. Questions 1 to 5 illustrate this question type. For each question, write the letter of the tool next to the number of the job described.

1 Measuring the diameter of a rod to be cut down to 0.726" on a lathe.
2 Measuring the spacing between two relay contacts required to be 0.005" apart.
3 Cutting 1 1/2" galvanized pipe quickly into many short sections.
4 Cutting 1 1/2" steel rod into short pieces.
5 Smoothing off a piece of pipe before cutting a thread.

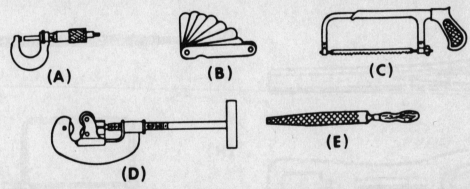

The correct tool for each of the operations indicated is as follows:

1. A (caliper) 3. D (pipe cutter) 5. E (file)
2. B (feeler gauge) 4. C (hack saw)

In Tool Analogy Tests, each question presents a numbered picture and four lettered pictures. The task is to select the one lettered tool or part that is most closely related to the numbered tool or part. Sample question 6 below illustrates this question type.

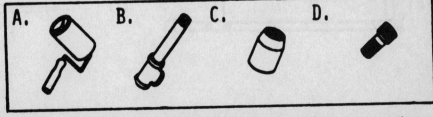

The correct answer to question 6 is B. The numbered picture shows a valve used to turn water on and off. The valve is most closely related to the water pipe pictured in alternative B.

Now go on to do as many of the practice questions as you can. Correct answers to all these questions will be found at the end of the chapter.

TEST I. TOOL RECOGNITION

DIRECTIONS: The items listed below are jobs, each of which normally requires the use of one of the tools or pieces of equipment shown. Read each item and, for the job given, select the required tool or piece of equipment. On your answer sheet write the letter of the tool in the blank provided next to the number of the job.

1. Drilling a 7/8-inch hole in a brick wall.
2. Tightening a water-pipe coupling.
3. Cutting a 1/2-inch stranded steel cable.
4. Tightening a lock nut on a surface mounted outlet box.
5. Doing scroll work on wood.
6. Rounding a hole in sheet metal.

7. Chipping off the corner of a brick.
8. Removing a 1/4-inch edge from a piece of sheet metal.
9. Making a hole to tap a thread in a steel block.
10. Converting a 4-foot brass rod into 1-foot pieces.
11. Making a hole in a door for a cylinder lock.

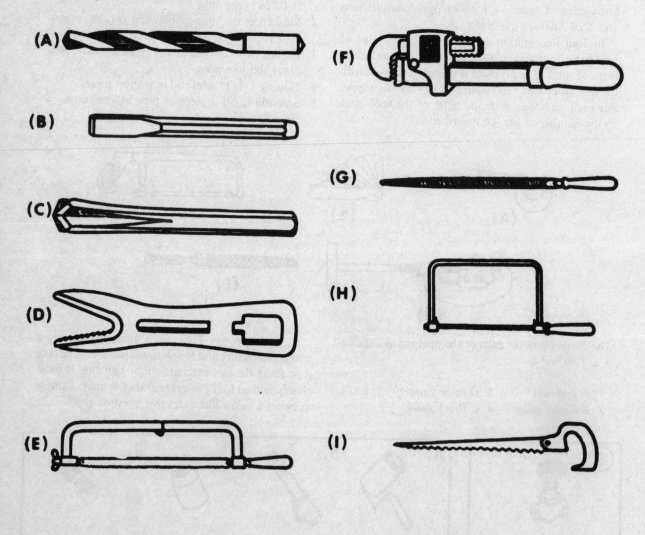

(A)

(B)

(C)

(D)

(E)

(F)

(G)

(H)

(I)

TEST II. TOOL RECOGNITION

DIRECTIONS: The items listed below are jobs, each of which normally requires the use of one of the tools shown below and on the next two pages. Read each item and, for the job given, select the proper tool to be used. On your answer sheet, write the letter of the tool next to the number of the job.

1. Laying out the angle of cut on the side supports for stair treads.
2. Cutting a brick accurately.
3. Marking lines parallel to the edge of a board.
4. Coupling nickel plated pipes.
5. Finishing the flat surface of a cement floor.
6. Checking that the side of a concrete form is vertical.
7. Drilling a 3/4″ hole in a 6″ channel iron on the job.
8. Locating the center of a circular plate having a diameter of 12″.
9. Forging an iron ring.
10. Bending a 1/2″ rod for a U-support for a 6″ pipe.
11. Coupling iron pipe for a handrail.
12. Chamfering the edge of a cement curb.
13. Filling joins in old brickwork.
14. Setting glazed tile on a wall.
15. Driving a drill for a hole through concrete for a 1″ pipe.
16. Grooving a cement floor to form blocks.
17. Laying out an accurate line for a saw cut as 90° to the edge of a board.
18. Holding sheet metal together for welding.
19. Locating a point for a hole in the floor directly under a point in the ceiling.
20. Cutting a channel in concrete for a pipe with the aid of tool "U".
21. Marking off a number of equal small distances.
22. Laying out a horizontal line in the center of a wall.
23. Driving a carpenter's chisel.
24. Driving a drill for a 1/4″ hole in a column.
25. Drilling holes in wood with auger bits.
26. Driving a screwdriver bit.
27. Making a chalk line.
28. Forcing a dove-tail joint together.
29. Bending thin sheet copper into the shape of a cylinder.
30. Marking a corner to be rounded on a board.
31. Flattening the end of a 1″ iron pipe on a steel plate.

TEST II

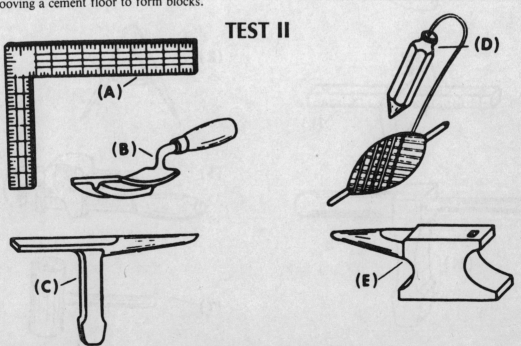

(A)

(B)

(C)

(D)

(E)

TEST II

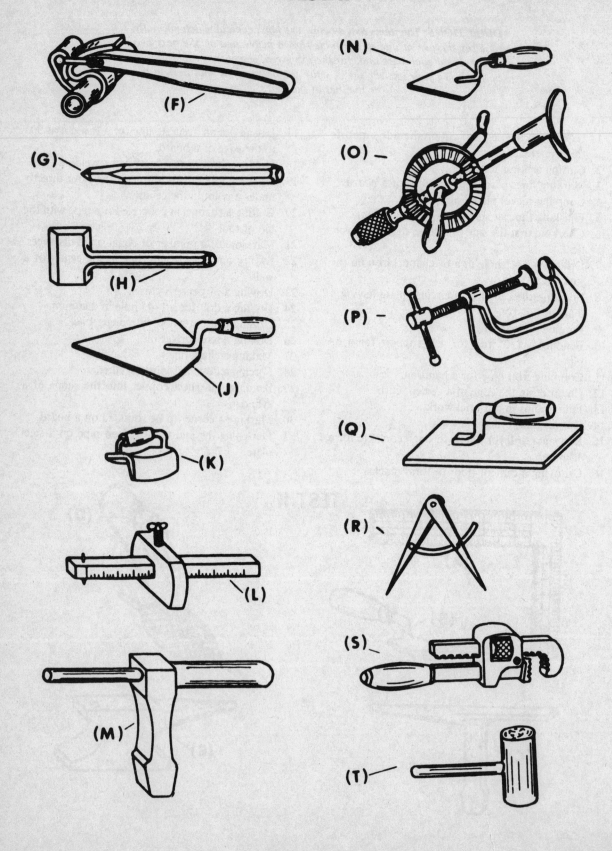

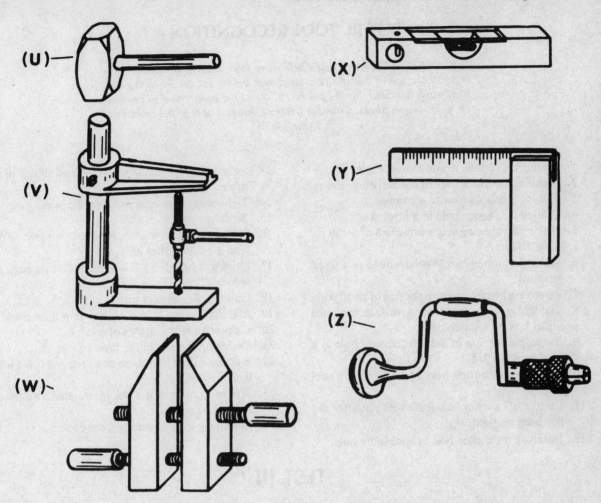

(U)

(V)

(W)

(X)

(Y)

(Z)

TEST III. TOOL RECOGNITION

DIRECTIONS: The items listed below are jobs, each of which normally requires the use of one of the tools shown below and on the next page. Read each item and, for the job given, select the proper tool to be used. On your answer sheet, write the letter of the tool next to the number of the job.

1. Cutting a 1/2-inch stranded steel cable.
2. Finishing the inside edge of conduit after cutting.
3. Turning a coupling on to a conduit.
4. Drilling a 3/4-inch hole in a brick wall.
5. Making an opening in the wood lath of a wall for a switch box.
6. Making a loop in No. 14 solid wire to fit a screw terminal.
7. Removing insulation from the end of small wire.
8. Tightening a lock nut on a surface mounted outlet box on a baseboard.
9. Increasing the size of a 1/8″ diameter hole in a steel beam to 3/8″.
10. Removing the button-head of a rivet set in a steel column.
11. Cutting off anchor bolt flush with nut after nut has been set tight.
12. Installing an anchor bolt in a masonry hole.

13. Sweating a ground clamp on the lead sheath of a cable.
14. Tightening a splice in No. 14 solid wire.
15. Bending conduit.
16. Drilling through a 14-inch hollow wooden wall with a bit 10-inches long.
17. Drilling a hole for 1 1/2-inch conduit through a wooden floor.
18. Cutting No. 18 steel wire.
19. Making a round hole oval shaped in sheet metal.
20. Chipping off the corner of a brick.
21. Widening the slot in the head of a screw.
22. Starting a tight thumb-screw to remove it from a light fixture.
23. Trimming off end of a No. 18 stranded wire close to a screw terminal.
24. Removing lead sheathing from cable.

TEST III

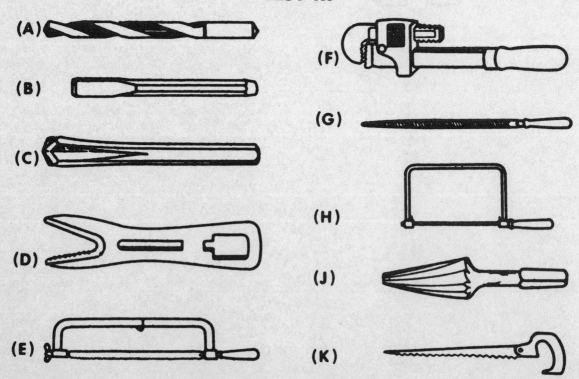

(A)

(B)

(C)

(D)

(E)

(F)

(G)

(H)

(J)

(K)

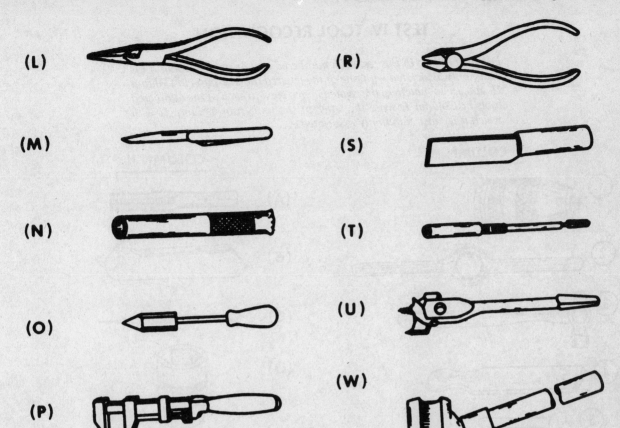

(L)

(M)

(N)

(O)

(P)

(R)

(S)

(T)

(U)

(W)

TEST IV. TOOL RECOGNITION

DIRECTIONS: In this test each numbered tool or part in Column I is commonly associated with one of the lettered tools or parts in Column II. You are to match up the related items. Next to each of the numbered tools in Column I, write the letter of the tool from Column II with which it is most commonly associated.

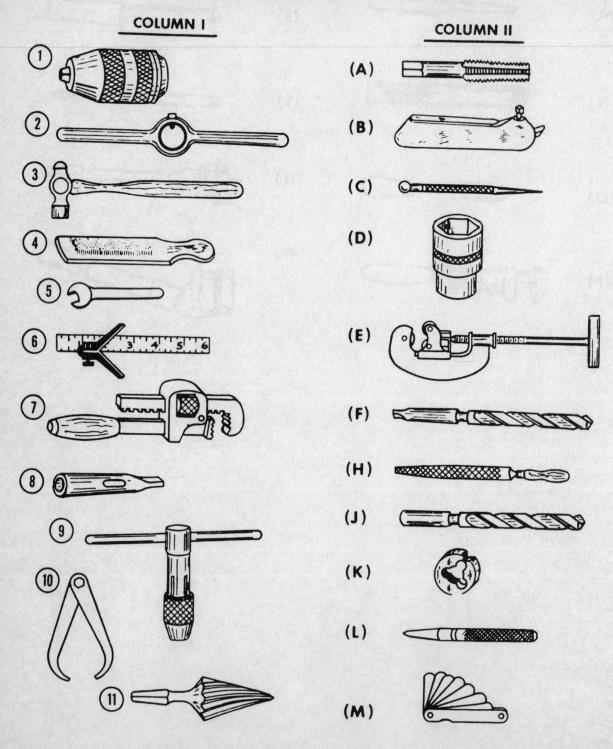

COLUMN I

COLUMN II

TEST V. TOOL RECOGNITION

DIRECTIONS: In this test, each numbered tool or part in Column I is commonly associated with one of the lettered tools or parts in Column II. You are to match up the related items. Next to the number of each tool in Column I, write the letter of the tool from Column II with which it is most commonly associated.

COLUMN I   **COLUMN II**

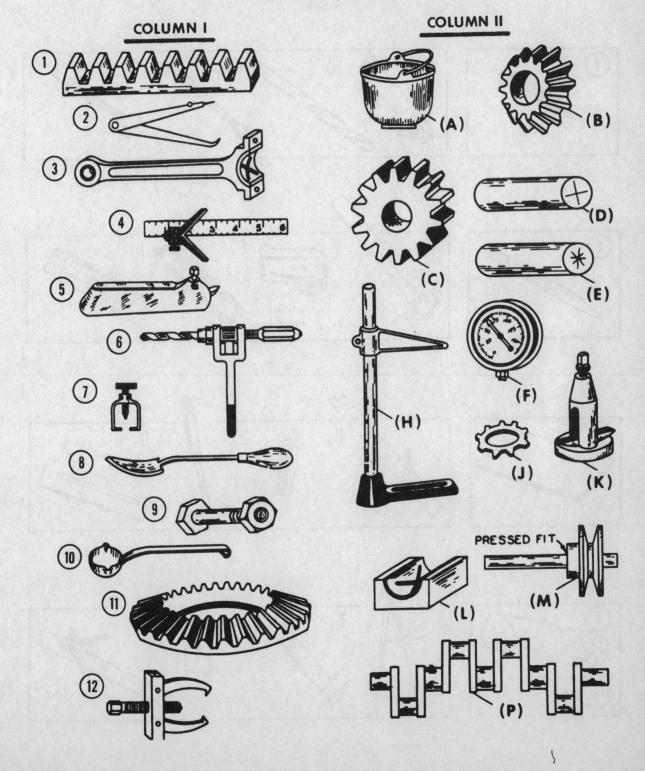

TEST VI. TOOL ANALOGY

DIRECTIONS: Each question in this test consists of a numbered picture followed by four lettered illustrations marked A, B, C, & D. The problem is to determine which of the four lettered pictures goes best with the numbered tool or machine part. For each question blacken the space on your answer sheet corresponding to the letter of the best answer.

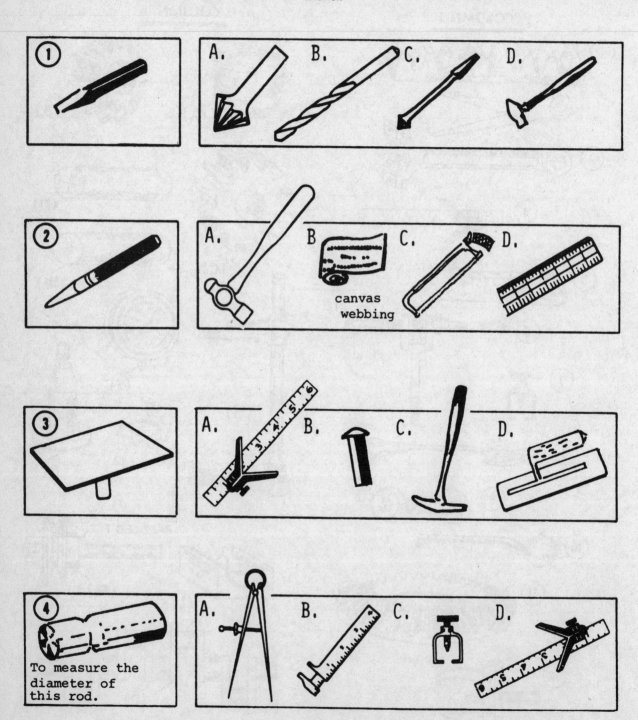

To measure the diameter of this rod.

A. 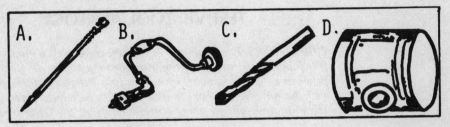 B. C. D.

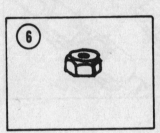

A. 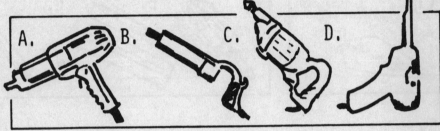 B. C. D.

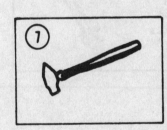

A. 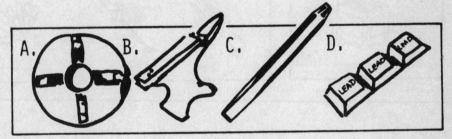 B. C. D.

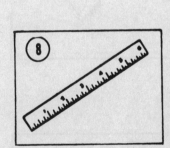

A. 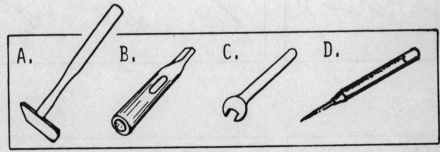 B. C. D.

TEST VII. TOOL ANALOGY

DIRECTIONS: Each question in this test consists of a numbered picture followed by four lettered illustrations marked A, B, C, & D. The problem is to determine which of the four lettered pictures goes best with the numbered tool or machine part. For each question blacken the space on your answer sheet corresponding to the letter of the best answer.

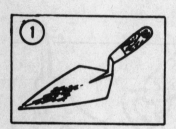
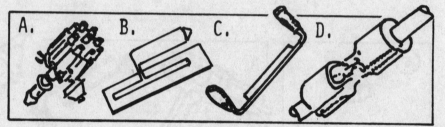

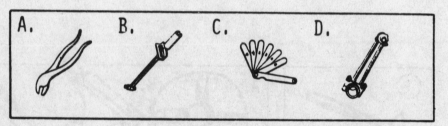

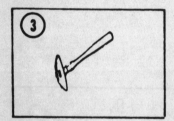
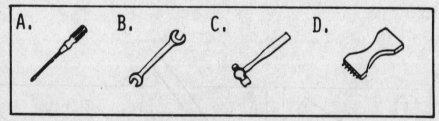

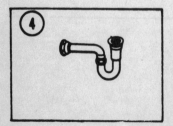
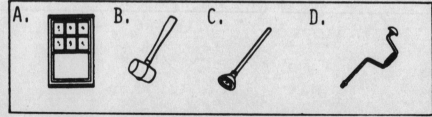

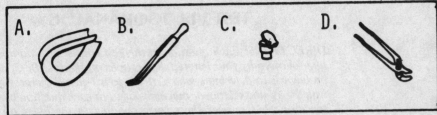

5 A. B. C. D.

6 A. B. C. D.

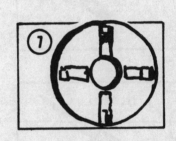

7 A. B. C. D.

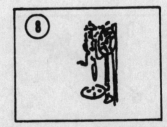
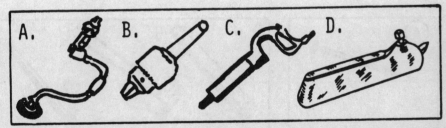

8 A. B. C. D.

TEST VIII. TOOL ANALOGY

DIRECTIONS: Each question in this test consist of a numbered picture followed by four lettered illustrations marked A, B, C, & D. The problem is to determine which of the four lettered pictures goes best with the numbered tool or machine part. For each question blacken the space on your answer sheet corresponding to the letter of the best answer.

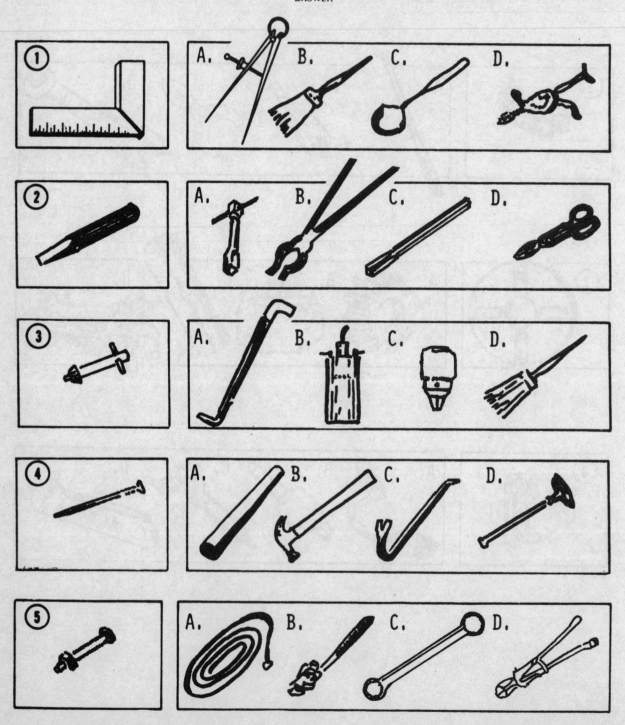

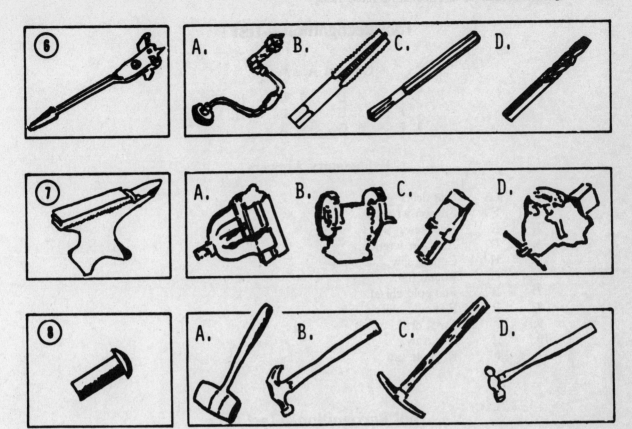

Tool Recognition—Test I

Correct Answers

1. C	4. D	7. B	10. E
2. F	5. H	8. B	11. I
3. E	6. G	9. A	

Explanatory Answers

1.	C	Star drill.
2.	F	Stillson wrench.
3.	E	Hacksaw.
4.	D	Alligator wrench.
5.	H	Coping saw.
6.	G	Round file.
7.	B	Flat cold chisel.
8.	B	Flat cold chisel.
9.	A	Twist drill.
10.	E	Hacksaw.
11.	I	Keyhole saw.

Tool Recognition—Test II

Correct Answers

1. A	9. E	17. Y	25. Z
2. H	10. E	18. P	26. Z
3. L	11. S	19. D	27. D
4. F	12. K	20. G	28. T
5. Q	13. N	21. R	29. M
6. X	14. J	22. X	30. R
7. V	15. U	23. T	31. U
8. A	16. B	24. O	

Explanatory Answers

1.	A	Carpenter's steel square.
2.	H	Brick chisel.
3.	L	Marking gauge.
4.	F	Strap wrench.
5.	Q	"Float" trowel.
6.	X	Level.
7.	V	Portable ratchet hand drillpress.
8.	A	Carpenter's steel square.
9.	E	Horn of the anvil.
10.	E	Horn of the anvil.
11.	S	Stillson pipe wrench.
12.	K	Edging tool.
13.	N	Mason's pointed trowel.
14.	J	Mason's brick trowel.
15.	U	Hand sledge hammer.

16.	B	Cement groover.
17.	Y	Try square.
18.	P	C-clamp.
19.	D	Plumb bob.
20.	G	Diamond nose chisel.
21.	R	Wing dividers.
22.	X	Level.
23.	T	Mallet.
24.	O	Breast drill.
25.	Z	Brace.
26.	Z	Brace.
27.	D	Chalk line usually used with plumb bob.
28.	T	Mallet.
29.	M	Mandrel.
30.	R	Wing dividers.
31.	U	Hand sledge hammer.

Tool Recognition—Test III

Correct Answers

1. E	7. M	13. O	19. G
2. J	8. D	14. L	20. B
3. F	9. A	15. W	21. E
4. C	10. B	16. T	22. L
5. K	11. E	17. U	23. R
6. L	12. N	18. R	24. S

Explanatory Answers

1.	E	Hacksaw.
2.	J	Tapered pipe reamer.
3.	F	Stillson pipe wrench.
4.	C	Star drill.
5.	K	Keyhole saw.
6.	L	Long nose pliers.
7.	M	Jacknife.
8.	D	Alligator wrench.
9.	A	Twist drill.
10.	B	Flat cold chisel.
11.	E	Hacksaw.
12.	N	Anchor driver.
13.	O	Soldering iron.
14.	L	Long nose pliers.
15.	W	Electrician's "hickey" (conduit bender).
16.	T	Extension bit holder.
17.	U	Expansive auger bit.
18.	R	Diagonal cutting pliers.
19.	G	Round file.
20.	B	Flat cold chisel.
21.	E	Hacksaw.
22.	L	Long nose pliers.
23.	R	Diagonal cutting pliers.
24.	S	Chisel knife.

Tool Recognition—Test IV

Correct Answers

1. J	4. H	7. E	10. B
2. K	5. B	8. F	11. E
3. L	6. C	9. A	

Explanatory Answers

1.	J	Drill chuck is used with twist drill (J).
2.	K	Die stock is used with round split die (K).
3.	L	Ball peen hammer is used with center punch (L).
4.	H	File card is used for cleaning teeth of file (H).
5.	B	Single end wrench used for the screw in tool holder (B).
6.	C	Center head and ruler used with scribe (C).
7.	E	Stillson wrench used to hold pipe being cut with pipe cutter (E).
8.	F	Tapered adapter sleeve used with tapered shank twist drill (F).
9.	A	T-handle tap wrench used with tap (A).
10.	B	Outside calipers used to measure diameter of part being cut in lathe with tool holder (B).
11.	E	Tapered pipe reamer used to chamfer inside edge of pipe cut with pipe cutter (E).

Tool Recognition—Test V

Correct Answers

1. C	5. K	9. J
2. E	6. H	10. A
3. P	7. F	11. B
4. D	8. L	12. M

Explanatory Answers

1.	C	Gear rack used with pinion gear (C).
2.	E	Hermaphrodite caliper used to locate center of rod (E).
3.	P	Piston crank operates crankshaft (P).
4.	D	Ruler with center head used to locate center of rod (D).
5.	K	Lathe tool holder held in tool post (K).
6.	H	Ratchet handle drill used with portable press (H).
7.	F	Tubing punch used to pierce hole in tubing for pressure gauge (F).
8.	L	Scraping tool used on bearing block (L).
9.	J	Nut and bolt used with lockwasher (J).
10.	A	Ladle used with solder pot (A).
11.	B	Large bevel gear used with small bevel gear (B).
12.	M	Wheel puller used to pull pulley (M) off shaft.

Tool Recognition—Test VI

Correct Answers

1. D	3. D	5. C	7. B
2. A	4. B	6. A	8. D

Explanatory Answers

1. D Flat cold chisel used with cross peen hammer (D).
2. A Center punch used with ball peen hammer (A).
3. D Mortar board would be used with trowel (D).
4. B Diameter of rod measured with side calipers (B).
5. C Drill chuck used to grip twist drill (C).
6. A Nut tightened/loosened with power nutdriver (A).
7. B Blacksmith's cross peen hammer used with anvil (B).
8. D Ruler used with scribe (D).

Tool Recognition—Test VII

Correct Answers

1. B	3. D	5. C	7. D
2. A	4. C	6. A	8. B

Explanatory Answers

1. B Steel finishing trowel (B) used to smooth mortar applied with pointed trowel.
2. A Cable connections on battery tightened with battery terminal pliers (A).
3. D Upholsterer's hammer used to tack webbing after it's stretched with spiked webbing stretcher (D).
4. C Drain unclogged with plunger (C).
5. C Bearings on motor have oil cups (C).
6. A Universal joint fitting and handle (A) are parts in a socket wrench set.
7. D Four-jaw independent chuck used to hold uneven stock in lathe (D).
8. B Drill press uses drill chuck (B) to hold drill.

Tool Recognition—Test VIII

Correct Answers

1. A	3. C	5. D	7. C
2. D	4. B	6. A	8. D

Explanatory Answers

1. A Graduated square used with dividers (A).
2. D Flat cold chisel and tin snips (D) are used in sheetmetal work.
3. C Chuck key used with drill chuck (C).
4. B Nail driven with claw hammer (B).
5. D Box wrench (D) used on nut or bolt.
6. A Expansive auger bit held in brace (A).
7. C Blacksmith's use a hardie (C) fitted into an anvil to notch or cut off stock.
8. D Rivets headed with a ball peen hammer (D).

MECHANICAL INSIGHT TESTS

The following questions have been selected from various civil service and private industry tests. All are designed to gauge your mechanical aptitude and your inherent feeling for machinery. They also measure your arithmetic ability.

HUNDREDS of civil service jobs and many jobs in private industry require the ability known as "mechanical insight"; in other words, the ability to visualize the operations of a machine in motion, to see the relationships among the different parts of a machine, and the capacity to make the necessary computations which are part of the job of a man or woman whose work is around machinery.

Many questions on the following pages are designed to test the candidate's ability to think in terms of a third dimension. Others deal with hydraulics, the forces exerted by fluids in a closed system, and with the workings of valves. Some call for knowledge of the operations of pulley systems and levers. In others, it is necessary to analyze the motion of interlocking gear systems.

You will find no "trick" or "catch" questions on a test of this type. The purpose of the examinations is to find the persons best qualified for work of a mechanical nature. Correct answers to all questions are consolidated at the end of this chapter.

A SAMPLE QUESTION EXPLAINED

In this and other mechanical insight questions, sharp observation and accurate interpretation of what is seen are of vital importance in arriving at the correct answer.

Upon careful observation, you can see that we are dealing with (1) a spring scale which registers weight when a force pulls down from below. (2) two 10-lb. weights; (3) a 10-lb bar. Once you have grasped the fact — through accurate observation — that this is a scale that registers weight when a force pulls from *below,* rather than a balance, you are well on your way to solving the question. If the drawing had represented a balance, the weight shown would, of course, have been zero, because the two equal 10-lb. weights on each side would have cancelled each other out, and the 10-lb bar itself would also not have registered because it is perfectly balanced.

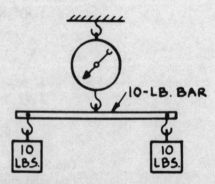

The reading on the weighing scale will be approximately

(A) zero   (C) 20 lbs.
(B) 10 lbs.   (D) 30 lbs.

The correct answer is (D) 30 lbs.

TEST I. MECHANICAL INSIGHT

DIRECTIONS: For each question read all the choices carefully. Then select that answer which you consider correct or most nearly correct. Blacken the answer space corresponding to your best choice, just as you would do on the actual examination.

1. Examine Figure 1 on the next page, and determine which of the following statements is true.
   (A) If the nut is held stationary and the head turned clockwise, the bolt will move up.
   (B) If the head of the bolt is held stationary and the nut is turned clockwise, the nut will move down.
   (C) If the head of the bolt is held stationary and the nut is turned clockwise, the nut will move up.
   (D) If the nut is held stationary and the bolt is turned counter-clockwise, the bolt will move down.

2. Referring to Figure 2, which one of the following statements is true?
   (A) If the nut is held stationary and the head turned clockwise, the bolt will move down.
   (B) If the head of the bolt is held stationary and the nut is turned clockwise, the nut will move down.
   (C) If the head of the bolt is held stationary and the nut is turned clockwise, the nut will move up.
   (D) If the nut is held stationary and the head turned counter-clockwise, the bolt will move up.

3. Figure 3 shows a bolt and nut and five numbered pieces. If all of the pieces are long enough to go through the bolt, and if the circular hole extends through the bolt and through the other side of the nut, which piece must you use to fix the nut in a stationary position?
   (A) 1   (B) 2   (E) 5
   (C) 3   (D) 4

4. Examine the tenon and the numbered mortises in Figure 4. The tenon best fits into the mortise numbered
   (A) 1   (B) 2   (E) 5
   (C) 3   (D) 4

5. In making the tenon in figure 4, the best of the following tools to use is
   (A) hammer   (B) knife
   (C) saw   (D) drill
                  (E) bit

6. Study the gear wheels in Figure 5, then determine which of the following statements is true.
   (A) If you turn wheel M clockwise by means of the handle, wheel P will also turn clockwise.
   (B) It will take the same time for a tooth of wheel P to make a full turn as it will for a tooth of wheel M.
   (C) It will take less time for a tooth of wheel P to make a full turn than it will take a tooth of wheel M.
   (D) It will take more time for a tooth of wheel P to make a full turn than it will for a tooth of wheel M.
   (E) The faster wheel P is turned, the slower wheel M will turn.

7. If wheel M in Figure 5 makes 16 full turns, the number of full turns made by wheel P will be
   (A) 20   (B) 12
   (C) 10   (D) 18

The Arrow Indicates a Clockwise Turn

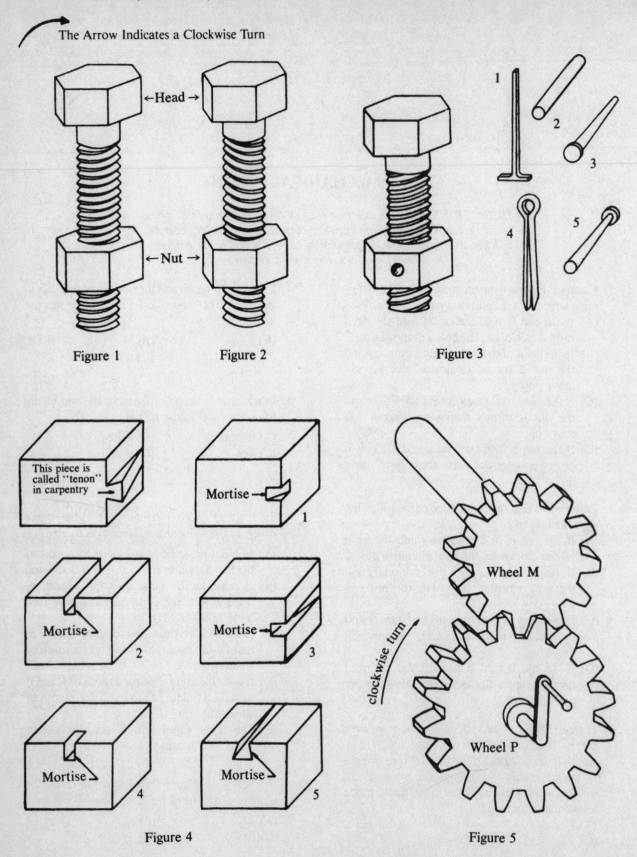

Figure 1

Figure 2

Figure 3

Figure 4

Figure 5

8. Referring to Figure 5, the number of teeth shown on wheel M is

(A) 12   (B) 14
(C) 16   (D) 10
           (E) 15

9. Referring to Figure 5, the number of teeth shown on wheel P is

(A) 10   (B) 18
(C) 16   (D) 17
           (E) 19

10. A bar measuring exactly three inches in length is pivoted at one end and a movement of 0.120 inches is noted at the opposite end. The movement of a point on the bar exactly 7/8″ from the pivot end will be

(A) .015 inches   (B) .105 inches
(C) .035 inches   (D) .35 inches

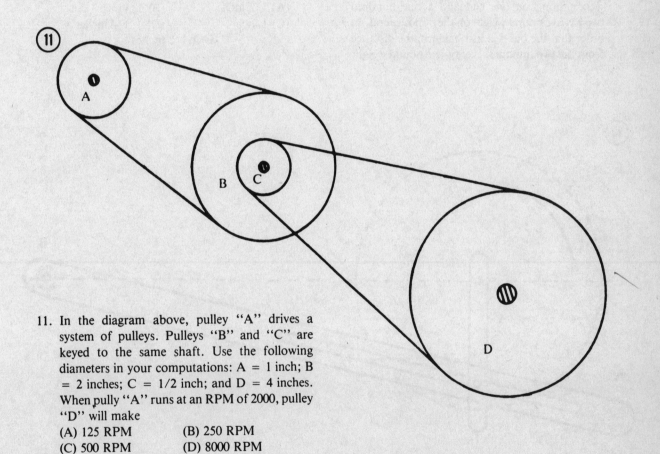

11. In the diagram above, pulley "A" drives a system of pulleys. Pulleys "B" and "C" are keyed to the same shaft. Use the following diameters in your computations: A = 1 inch; B = 2 inches; C = 1/2 inch; and D = 4 inches. When pully "A" runs at an RPM of 2000, pulley "D" will make

(A) 125 RPM   (B) 250 RPM
(C) 500 RPM   (D) 8000 RPM

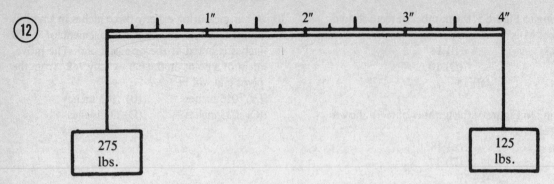

12. The bar above, which is exactly four inches in length, has a two hundred seventy-five pound weight hung on one end and a one hundred twenty-five pound weight on the opposite end. In order that the bar will just balance, the distance from the two hundred seventy-five pound weight to the fulcrum point should be (In your computation neglect the weight of the bar.)

(A) 1/2 inch   (B) 3/4 inch
(C) 1 inch   (D) 1 1/4 inches
       (E) 1-1/2 inches

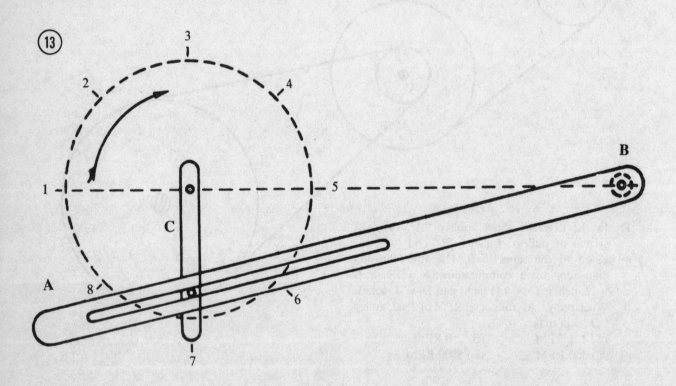

13. In the diagram above, crank arm "C" revolves at a constant speed of 400 RPM and drives the lever "AB". When lever "AB" is moving the fastest arm "C" will be in position

(A) 1   (B) 5
(C) 6   (D) 7

14. In the diagram shown, the axle eight inches in diameter has attached a handle 28 inches in diameter. If a force of 50 lbs. is applied to the handle, the axle will lift a weight of
(A) 224 lbs.   (B) 200 lbs.
(C) 175 lbs.   (D) 88 lbs.
       (E) 75 lbs.

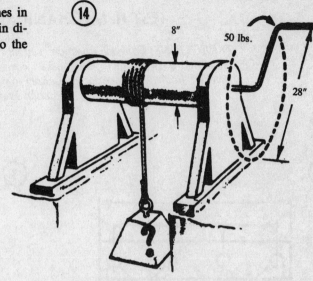

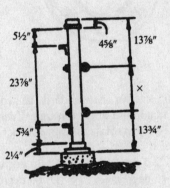

15. On the post, the dimension marked "X" is
(A) 9 3/4"   (B) 10 3/4"
(C) 13 3/8"   (D) 14 3/8"

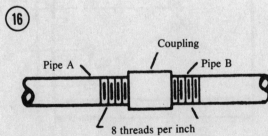

16. If pipe A is held in a vise and pipe B is turned ten revolutions with a wrench, the overall length of the pipes and coupling will decrease
(A) 5/8 inch   (B) 1 1/4 inches
(C) 2 1/2 inches   (D) 3 3/4 inches

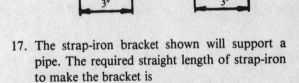

17. The strap-iron bracket shown will support a pipe. The required straight length of strap-iron to make the bracket is
(A) 20 1/2 inches   (B) 17 inches
(C) 15 inches   (D) 13 1/4 inches

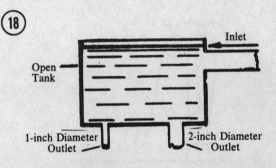

18. Eight gallons per minute of water flow at a given time from the one inch outlet in the tank shown. What is the amount of water flowing at that time from the two-inch outlet?
(A) 64 gallons per minute
(B) 32 gallons per minute
(C) 16 gallons per minute
(D) 2 gallons per minute

TEST II. MECHANICAL INSIGHT

DIRECTIONS: For each question read all the choices carefully. Then select that answer which you consider correct or most nearly correct. Blacken the answer space corresponding to your best choice, just as you would do on the actual examination.

①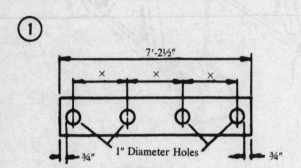

1. The dimension "X" on the piece shown is
   (A) 2' - 3 2/3" (B) 2' -4"
   (C) 2' - 4 1/3" (D) 2' - 5 1/4"

③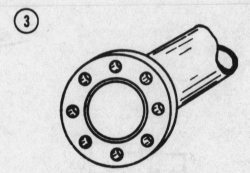

3. In the case of the standard flanged pipe shown, the maximum angle through which it would be necessary to rotate the pipe in order to line up the holes is
   (A) 22.5 degrees (B) 30 degrees
   (C) 45 degrees (D) 60 degrees

②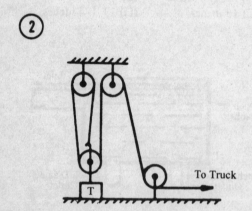

2. The tank "T" is to be raised as shown by attaching the pull rope to a truck. If the tank is to be raised ten feet, the truck will have to move
   (A) 20 feet (B) 30 feet
   (C) 40 feet (D) 50 feet

④

4. The distance "X" from center to center of the two holes is
   (A) 10 inches (B) 9 inches
   (C) 8 1/2 inches (D) 6 inches

⑤

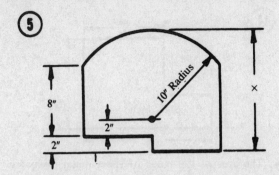

5. The distance "X" on the piece shown is
   (A) 16 inches (B) 14 inches
   (C) 12 inches (D) 10 inches

⑥

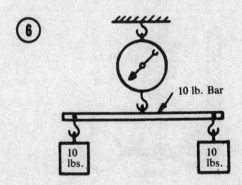

6. The reading on the weighing scale will be approximately
   (A) zero (B) 10 lbs.
   (C) 20 lbs. (D) 30 lbs.

⑦

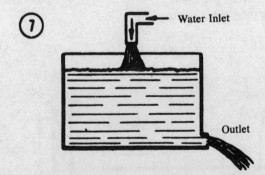

7. If water is flowing into the tank at the rate of 120 gallons per hour and flowing out of the tank at a constant rate of one gallon per minute, the water level in the tank will
   (A) rise 1 gallon per minute
   (B) rise 2 gallons per minute
   (C) fall 2 gallons per minute
   (D) fall 1 gallon per minute

⑧

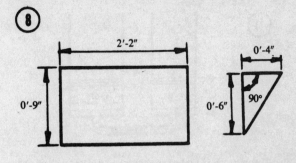

8. The maximum number of triangular pieces shown which can be cut from the piece of sheet metal shown is
   (A) 12 (B) 16
   (C) 20 (D) 25

⑨

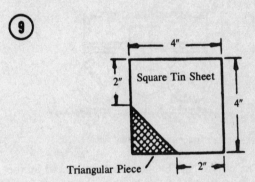

9. The maximum number of triangular pieces which can be cut from the tin sheet is
   (A) 10 (B) 8
   (C) 6 (D) 4

⑩

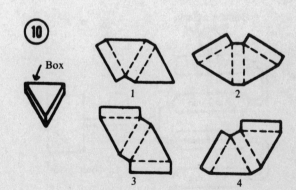

10. The flat sheet metal pattern which can be bent along the dotted lines to form the completely closed triangular box is
    (A) 1 (B) 2
    (C) 3 (D) 4

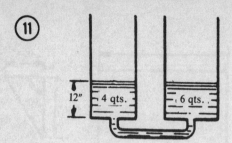

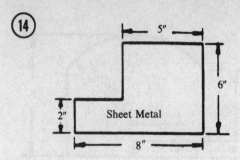

11. To bring the level of the water in tanks to a height of 2 1/2 feet, the quantity of water to be added is
    (A) 10 qts.   (B) 15 qts.
    (C) 20 qts.   (D) 25 qts.

14. The area of the piece of sheet metal in square inches is
    (A) 48 in.²   (B) 36 in.²
    (C) 20 in.²   (D) 16 in.²

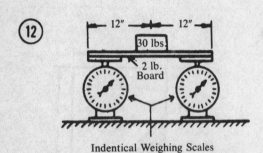

Indentical Weighing Scales

12. The weight held by the board and placed on the two identical scales will cause *each* scale to read
    (A) 8 lbs.   (B) 15 lbs.
    (C) 16 lbs.   (D) 32 lbs.

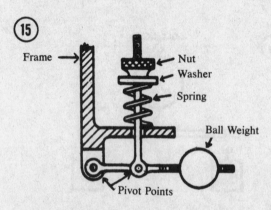

15. If the ball and spring mechanism are balanced in the position shown, the ball will move upward if
    (A) the nut is loosened
    (B) ball is moved away from the frame
    (C) the nut is loosened and the ball moved away from the frame
    (D) the nut is tightened

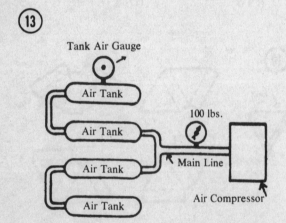

13. Four air reservoirs have been filled with air by the air compressor. If the main line air gauge reads 100 lbs. then the tank air gauge will read
    (A) 25 lbs.   (B) 50 lbs.
    (C) 100 lbs.   (D) 200 lbs.

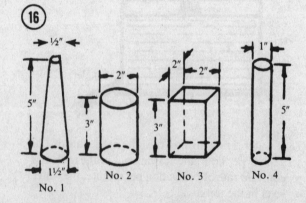

16. The container which will hold the most water is
    (A) No.1   (B) No. 2
    (C) No. 3   (D) No. 4

TEST III. MECHANICAL INSIGHT

DIRECTIONS: For each question read all the choices carefully. Then select that answer which you consider correct or most nearly correct. Blacken the answer space corresponding to your best choice, just as you would do on the actual examination.

Items 1 to 10 inclusive refer to the following, boxed figures. Each item gives the proper figure to use with that item.

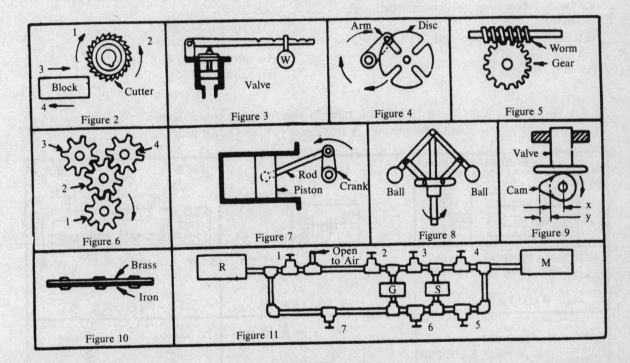

1. Figure 2 shows a cutter and a steel block. For proper cutting, they should move respectively in directions
   (A) 1 and 4   (B) 1 and 3
   (C) 2 and 3   (D) 2 and 4

2. Figure 3 shows a lever type safety valve. It will blow off at a lower pressure if weight W is
   (A) increased
   (B) moved to the right
   (C) increased and moved to the right
   (D) moved left

3. Figure 4 shows a slotted disc turned by a pin on a rotating arm. One revolution of the arm turns the disc
   (A) 1/4 turn   (B) 3/4 turn
   (C) 1/2 turn   (D) one complete turn

4. Figure 5 shows a worm and a gear. If the worm rotates slowly on its shaft, the gear will
   (A) not turn   (B) turn rapidly
   (C) turn very slowly   (D) oscillate

5. Figure 6 shows four gears. If gear 1 turns as shown, then the gears turning in the same direction are
(A) 2, 3 and 4   (B) 2 and 3
(C) 2 and 4   (D) 3 and 4

6. Figure 7 shows a crank and piston. The piston moves from mid-position to the extreme right, if the crank
(A) makes 1/2 turn
(B) makes a 3/4 turn
(C) makes one turn
(D) eakes 1 1/2 turns

7. Figures 8 shows a governor on a rotating shaft. As the shaft speeds up, the governor balls will
(A) move down
(B)) move upward and inward
(C) move upward
(D) move inward

8. Figure 9 shows a cam and a valve. For each cam revolution, the vertical valve rise equals distance
(A) Y   (B) X plus Y
(C) X   (D) twice X

9. Figure 10 shows a brass and iron strip continuously riveted together. High temperatures would probably
(A) have no effect at all
(B) bend the strips
(C) separate the strips
(D) shorten the strips

10. In Figure 11, assume all valves closed. For air flow from R, through G, then through S, to M open
(A) valves 1, 2, 6 and 4
(B) valves 7, 3 and 4
(C) valves 7, 6 and 4
(D) valves 7, 3 and 5

Items 11 to 16 inclusive refer to the following boxed figures. Each item gives the proper figure to use with that item.

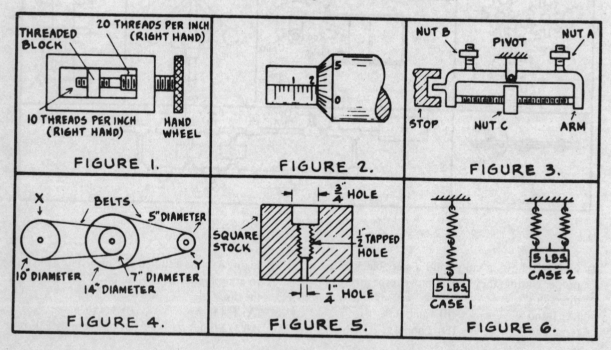

11. In Figure 1, the threaded block can slide in the slot but cannot revolve. If the hand wheel is turned 20 revolutions clockwise, the threaded block will move
(A) one inch to the left
(B) 1/2 inch to the left

(C) one inch to the right
(D) 1/2 inch to the right

12. The micrometer in Figure 2 reads
(A) .2270   (C) .2120
(B) .2252   (D) .2020

13. The arm in Figure 3 is exactly balanced as shown. If nut "A" is removed entirely then, in order to rebalance the arm, it will be necessary to turn
    (A) nut "C" toward the right
    (B) nut "C" toward the left
    (C) nut "B" up
    (D) nut "B" down

14. A double belt drive is shown in Figure 4. If the pulley marked "X" is revolving at 100 RPM, the speed of pulley "Y" is
    (A) 800 RPM   (B) 400 RPM
    (C) 200 RPM   (D) 25 RPM

15. To drill and tap the holes in Figure 5, the best practice would be to do the work in the following order

    (A) drill 1/2" hole, drill 3/4" hole, tap 1/2" hole, drill 1/4" hole from bottom
    (B) drill 1/4" hole through piece, drill 1/2" hole, tap 1/2" hole, drill 3/4" hole
    (C) drill 3/4" hole, drill the 1/2" hole, drill the 1/4" hole, tap the 1/2" hole
    (D) drill 1/4" hole through piece, drill 1/2" hole, drill 3/4" hole, tap 1/2" hole

16. In Figure 6, all 4 springs are identical, In Case 1 with the springs end to end, the stretch of each spring due to the five lb. weight is
    (A) 1/2 as much as in Case 2
    (B) the same as in Case 2
    (C) twice as much as in Case 2
    (D) four times as much as in Case 2

TEST IV. MECHANICAL INSIGHT

DIRECTIONS: For each of the following questions, select the choice which best answers the question or completes the statement.

①

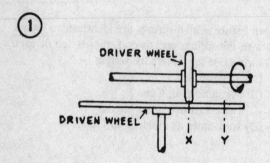

1. When the driver wheel is moved from location X to location Y, the driven wheel will
   (A) reverse its direction of rotation
   (B) turn slower
   (C) not change its speed of rotation
   (D) turn faster

③

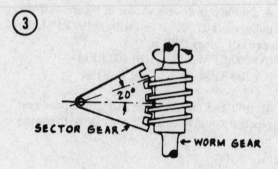

3. One revolution of the worm gear will turn the sector gear through the angle of
   (A) 30°   (B) 20°
   (C) 10°   (D) 5°

②

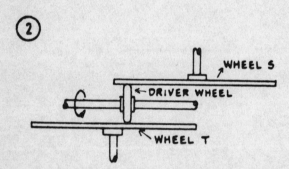

2. With the wheels in the position shown
   (A) wheels S and T will rotate in opposite directions
   (B) wheels S and T will rotate at the same speed
   (C) wheels S and T will rotate in the same direction
   (D) wheel S will rotate at exactly the same speed as the driver wheel

④

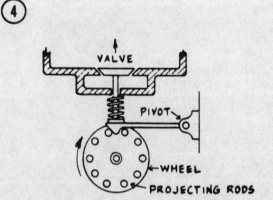

4. In order to open the valve once every second the wheel must rotate at
   (A) 30 RPM   (B) 20 RPM
   (C) 10 RPM   (D) 6 RPM

⑤

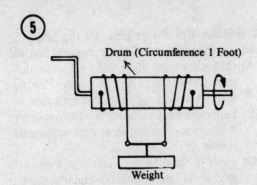

Drum (Circumference 1 Foot)

Weight

5. One complete revolution of the windlass drum will move the weight up
   (A) 1/2 foot   (B) 1 foot
   (C) 1 1/2 feet   (D) 2 feet

⑥

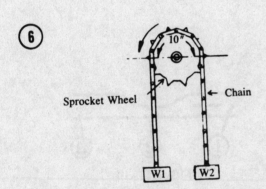

Sprocket Wheel   ← Chain

W1   W2

6. One complete revolution of the sprocket wheel will bring weight W2 higher than weight W1 by
   (A) 20″   (B) 30″
   (C) 40″   (D) 50″

⑦

Metal Plate

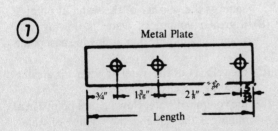

Length

7. The total length of the metal plate shown is
   (A) 3 7/32 inches   (B) 3 9/32 inches
   (C) 4 7/32 inches   (D) 4 9/32 inches

⑧

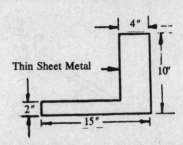

Thin Sheet Metal

8. The maximum number of rectangular pieces, each two inches by eight inches which can be cut from the thin metal sheets shown is
   (A) One   (B) Two
   (C) Three   (D) Four

⑨

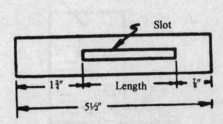

Slot

9. The length of the slot in the piece shown is
   (A) 2 7/8 inches   (B) 2 5/8 inches
   (C) 2 3/8 inches   (D) 2 1/8 inches

⑩

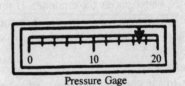

Pressure Gage

10. The reading shown on the gage is
    (A) 10.35   (B) 10.7
    (C) 13.5   (D) 17.0

⑪

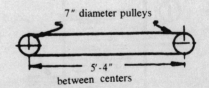

7″ diameter pulleys

5′-4″ between centers

11. The total length of the belt connecting the two pulleys is
    (A) 7 feet, 2 inches   (B) 9 feet, 0 inches
    (C) 12 feet, 6 inches   (D) 14 feet, 4 inches

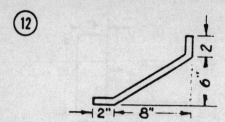

12. The length of the strap before bending is
    (A) 18 inches   (B) 14 inches
    (C) 13 inches   (D) 11 inches

FIGURE I.

13. Figure I represents an enclosed water chamber, partially filled with water. The number 1 indicates air in the chamber and 2 indicates a pipe by which water enters the chamber. If the water pressure in the pipe, 2, increases then the
    (A) water pressure in the chamber will be decreased
    (B) water level in the chamber will fall
    (C) air in the chamber will be compressed
    (D) air in the chamber will expand
    (E) water will flow out of the chamber

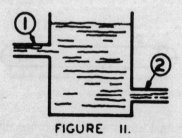

FIGURE II.

14. Figure II represents a water tank containing water. The number 1 indicates an intake pipe and

2 indicates a discharge pipe. Of the following, the statement which is least accurate is that the
(A) tank will eventually overflow if water flows through the intake pipe at a faster rate than it flows out through the discharge pipe
(B) tank will empty completely if the intake pipe is closed and the discharge pipe is allowed to remain open
(C) water in the tank will remain at a constant level if the rate of intake is equal to the rate of discharge
(D) water in the tank will rise if the intake pipe is operating when the discharge pipe is closed
(E) time required to fill the tank, if the discharge pipe is closed, depends upon the rate of flow of water through the intake pipe

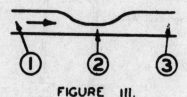

FIGURE III.

15. Figure III represents a pipe through which water is flowing in the direction of the arrow. There is a constriction in the pipe at the point indicated by the number 2. Water is being pumped into the pipe at a constant rate of 350 gallons per minute. Of the following, the most accurate statement is that
(A) the velocity of the water at point 2 is the same as the velocity of the water at point 3
(B) a greater volume of water is flowing past point 1 in a minute than is flowing past point 2
(C) the velocity of the water at point 1 is greater than the velocity at pont 2
(D) the volume of water flowing past point 2 in a minute is the same as the volume of water flowing past point 1 in a minute
(E) a greater volume of water is flowing past point 3 in a minute than is flowing past point 2

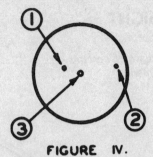

FIGURE IV.

16. Figure IV represents a revolving wheel. The numbers 1 and 2 indicate two fixed points on the wheel. The number 3 indicates the center of the wheel. Of the following, the most accurate statement is that
(A) point 1 makes fewer revolutions per minute than point 2
(B) point 2 makes more revolutions per minute than point 1
(C) point 2 traverses a greater linear distance than point 1
(D) point 1 will make a complete revolution in less time than point 2
(E) the product of the linear distance traversed by either point and the time required for one revolution is equal to the number of revolutions

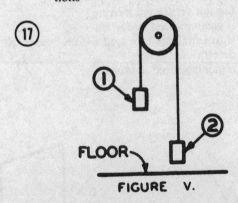

FLOOR

FIGURE V.

17. Figure V represents a pulley, with practically no friction, from which two ten pound weights are suspended as indicated. If a downward force is applied to weight 1, it is most likely that weight 1 will
(A) come to rest at the present level of weight 2
(B) move downward until it is level with weight 2
(C) move downward until it reaches the floor
(D) pass weight 2 in its downward motion and then return to its present position
(E) move downward a short distance before the direction of movement is reversed

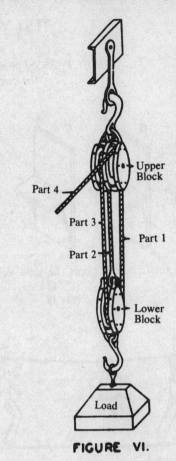

FIGURE VI.

18. Examine Figure VI and determine which part of the rope is fastened directly to the block.
(A) Part I   (B) Part 2
(C) Part 3   (D) Part 4
(E) no part

19. If you study Figure VI you will find that only one of the following statements is true:
(A) Each part of the rope carries about the same share of the load.
(B) A load cannot be lowered by means of the tackle shown.
(C) If the upper block were smaller but its weight unchanged, hoisting would be easier for the operator.
(D) The rope cannot possibly be connected to a winch.
(E) If the lower block were made smaller but its weight kept the same, hoisting would be easier for the operator

TEST V. MECHANICAL INSIGHT

DIRECTIONS: For each question, select the choice which best answers the question or completes the statement.

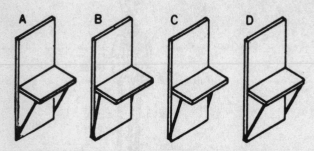

1. Which shelf could support the most weight?
   (A) A
   (B) B
   (C) C
   (D) D

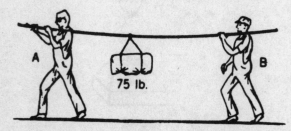

2. The weight is being carried entirely on the shoulders of the two men shown. Which man bears the most weight on his shoulder?
   (A) A
   (B) B
   (C) Both men are carrying the same.
   (D) It is impossible to tell.

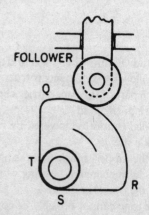

3. The follower is at its highest position between points
   (A) Q and R
   (B) R and S
   (C) S and T
   (D) T and Q

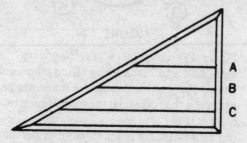

4. All of the wires are of the same substance, the same diameter, and under the same tension. Which will vibrate at the highest frequency?
   (A) A
   (B) B
   (C) C
   (D) They will vibrate at equal frequency

5. A man in an elevator is carrying a heavy suitcase. The suitcase will feel heaviest to him when the elevator
   (A) has not yet started moving
   (B) is gaining speed in descent
   (C) is maintaining a rapid steady speed of descent
   (D) is gaining speed in ascent

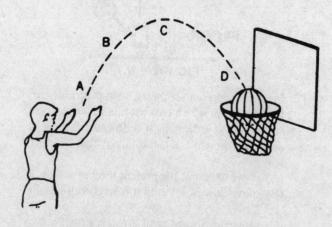

6. At which point was the basketball moving slowest?
   (A) A
   (B) B
   (C) C
   (D) D

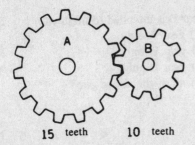

15 teeth   10 teeth

7. If gear A makes 14 revolutions, gear B will make
   (A) 21
   (B) 17
   (C) 14
   (D) 9

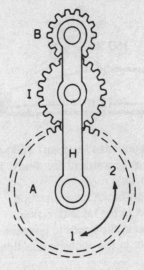

10. If arm H is held fixed as gear B turns in direction
    2, gear
    (A) A must turn in direction 1.
    (B) A must turn in direction 2.
    (C) I must turn in direction 2.
    (D) A must be held fixed.

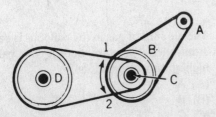

8. If pulley A is the driver and turns in direction 1,
   which pulley turns fastest?
   (A) A
   (B) B
   (C) C
   (D) D

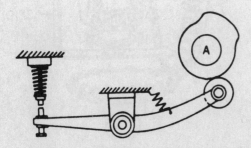

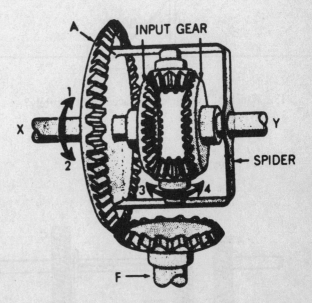

9. As cam A makes one complete turn, the setscrew
   will hit the contact point
   (A) once
   (B) twice
   (C) three times
   (D) not at all

11. If shaft X turns in direction 2 as shaft Y is held
    fixed, shaft F will turn in direction
    (A) 3 and gear A in direction 1.
    (B) 3 and gear A in direction 2.
    (C) 4 and gear A in direction 1.
    (D) 4 and gear A in direction 2.

12. A 150-pound man jumps off a 600-pound raft to a point in the water 12 feet away. Theoretically, the raft would move
(A) 12 feet in the same direction.
(B) 6 feet in the same direction.
(C) 3 feet in the opposite direction.
(D) 1 foot in the opposite direction.

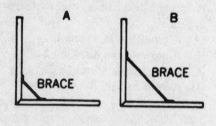

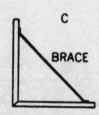

13. Which of the angles is braced most securely?
(A) A
(B) B
(C) C
(D) All equally braced.

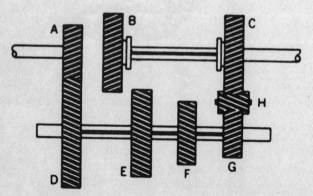

14. Gear B is intended to mesh with
(A) gear A only
(B) gear D only
(C) gear E only
(D) all of the above gears

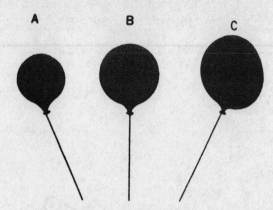

15. The amount of gas in the balloons is equal. The atmospheric pressure outside the balloons is highest on which balloon?
(A) A
(B) B
(C) C
(D) The pressure is equal on all balloons.

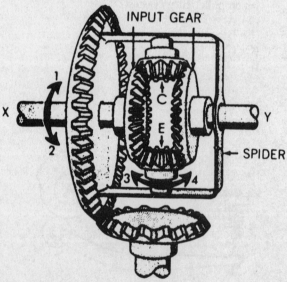

16. If shaft X turns in direction 1 and shaft Y is held fixed, gear C will turn in direction
(A) 3 and gear E in direction 3
(B) 3 and gear E in direction 4
(C) 4 and gear E in direction 3
(D) 4 and gear E in direction 4

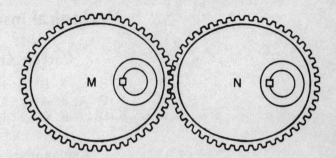

19. If gear N turns at a constant RPM, gear M turns at
    (A) the same constant RPM as N
    (B) a faster constant RPM than N
    (C) a slow constant RPM than N
    (D) a variable RPM

17. Liquid is being transferred from the barrel to the bucket by
    (A) suction in the hose
    (B) fluid pressure in the hose
    (C) air pressure on top of the liquid
    (D) capillary action

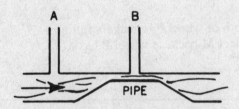

18. If water is pumped rapidly through the pipe in the direction shown by the arrow, it will
    (A) rise higher in tube A than in tube B
    (B) rise higher in tube B than in tube A
    (C) rise in tube A but not in tube B
    (D) rise in tube B but not in tube A

20. If gear M turns at a constant RPM gear N turns at a number of RPM that
    (A) is constant and less than that of M
    (B) is constant and the same as M
    (C) reaches its maximum four times each revolution
    (D) reaches its maximum eight times each revolution

Mechanical Insight—Test I

Correct Answers

1. B	4. E	7. B	10. C	13. B	16. B
2. C	5. C	8. A	11. A	14. C	17. C
3. D	6. D	9. C	12. D	15. D	18. B

Explanatory Answers

1. B If the bolt is held stationary and the nut is turned clockwise, the nut will move down because of the right-hand thread.

2. C If the head of the bolt is held stationary and the nut is turned clockwise, the nut will move up because of the left-hand thread.

3. D Cotter pin (#4) would fit through the circular hole in the nut and bolt to fix a stationary position.

4. E Mortise (#5) is the mating shape to the tenon.

5. C Saw (C) is most suitable for cutting away material to form the tenon.

6. D It will take more time for a tooth of wheel P to make a full turn than it would for a tooth of wheel M because wheel P has a larger diameter.

7. B Inversely Proportional.

$$\underbrace{\text{(teeth) (turns)}}_{P} = \underbrace{\text{(teeth) (turns)}}_{M}$$

$$(16)\,(x) = (12)\,(16)$$
$$x = 12 \text{ turns}$$

8. A Self-explanatory. See Figure 5.

9. C Self-explanatory. See Figure 5.

10. C

$$\frac{.120}{3} = \frac{x}{.875}$$

$$x = .035 \text{ inches}$$

11.　A　Inversely proportional.

$$\overbrace{(RPM)\ (diameter)}^{A} = \overbrace{(RPM)\ (diameter)}^{B}$$

$$(2000)\ (1) = x(2)$$
$$x = 1000\ RPM$$

C runs at the same RPM as B

$$\underbrace{(RPM)\ (diameter)}_{C} = \underbrace{(RPM)\ (diameter)}_{D}$$

$$(1000)\ (\tfrac{1}{2}) = x(4)$$
$$x = 125\ RPM$$

12.　D

$$(275)\ (x) = (4 - x)\ (125)$$
$$275x = 500 - 125x$$
$$400x = 500$$
$$x = 1.25$$
$$x = 1\tfrac{1}{4}\ inches$$

13.　B　At position 5, the pin on the crank arm (fulcrum) is closest to pivot B.

14.　C　

$$(force)\ (distance) = (force)\ (distance)$$
$$(50)\ (28) = x(8)$$
$$x = 175\ lbs.$$

15.　D

$$5\tfrac{1}{2} + 23\tfrac{7}{8} + 5\tfrac{3}{4} + 2\tfrac{1}{4}$$
$$= 37\tfrac{3}{8}$$

$$13\tfrac{7}{8} + x + 13\tfrac{3}{4} = 27\tfrac{5}{8} + x$$

$$(27\tfrac{5}{8} + x) - 4\tfrac{5}{8} = 37\tfrac{3}{8}$$
$$23 + x = 37\tfrac{3}{8}$$
$$x = 14\tfrac{3}{8}''$$

16.　B　$10(\tfrac{1}{8}) = \tfrac{10}{8} = 1\tfrac{1}{4} = 1\tfrac{1}{4}$ inches

17.　C

$$Length = 3 + 1\tfrac{3}{4} + \tfrac{1}{2}(2\pi r) + 1\tfrac{3}{4} + 3$$
$$Length = 4\tfrac{3}{4} + (\tfrac{22}{7})(\tfrac{7}{4}) + 4\tfrac{3}{4}$$
$$Length = 15\ inches$$

18.　B

$$\frac{rate}{area} \qquad \frac{8}{\frac{\pi}{4}} = \frac{x}{\pi}$$

$$x = 32\ gallons\ per\ minute$$

$$A = \pi(\tfrac{1}{2})^2 \qquad A = \pi(1)^2$$
$$A = \frac{\pi}{4} \qquad\qquad = \pi$$

Mechanical Insight—Test II

Correct Answers

1. B	3. A	5. B	7. A	9. B	11. B	13. C	15. D
2. B	4. A	6. D	8. B	10. C	12. C	14. B	16. C

Explanatory Answers

1. B

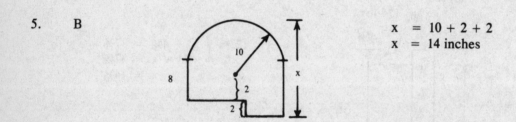

$2(\frac{3}{4}) + 2(\frac{1}{2}) + 3x = 86.5$ in.

$2.5 + 3x = 86.5$ in.

$3x = 84$ in.

$x = 28$ in. $= 2$ ft.-4 in.

$x = 2'\text{-}4''$

2. B   Three supporting ropes. IMA = 3, therefore 10(3) = 30 feet.

3. A $\frac{360}{8} = 45°$ between holes. $22\frac{1}{2}°$ is the furthest it would be necessary to rotate the pipe.

4. A

$x^2 = 6^2 + 8^2$

$x^2 = 100$

$x = 10$ inches

5. B

$x = 10 + 2 + 2$

$x = 14$ inches

6. D

scale reads 30 lbs.

7. A   in: $\dfrac{120 \text{ gal.}}{\text{hr.}} = \dfrac{2 \text{ gal.}}{\text{min.}}$ ⎫

out: $\dfrac{1 \text{ gal.}}{\text{min.}}$ ⎬ net: rise of one gallon per minute

⎭

8. B

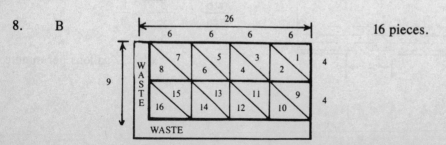

16 pieces.

9. B   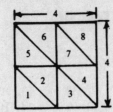   8 pieces.

10. C   Self-explanatory.

11. B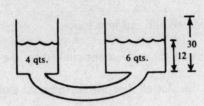

$$\frac{12 \text{ in.}}{4 \text{ qts.} + 6 \text{ qts.}} = \frac{18 \text{ in. more}}{x}$$

$$\frac{12}{10} = \frac{18}{x}$$

$$x = 15 \text{ qts.}$$

12. C   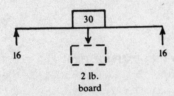   Each scale reads 16 lbs.

13. C   They all have the same pressure, causing the tank air gauge to read 100 lbs.

14. B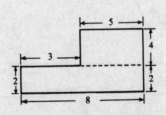

$$A = 2(8) + 5(4)$$
$$A = 36 \text{ in.}^2$$

15. D   Self-explanatory. See diagram.

16. C $V_{\text{cyl.}} = \pi r^2 h$; $V_{\text{rect.}} = lwh$

(A)   Use avg. radius = ½
$$\approx \pi(\tfrac{1}{2})^2(5) = \frac{5\pi}{4} \approx 9''$$

(B) $V = \pi(^2/_2)^2(3)$
$$= 3\pi \approx 9.4''$$

(C) $V = 3(2)(2)$
$$= 12''$$
    LARGEST

(D) $V = \pi(\tfrac{1}{2})^2(5)$
$$= \frac{5\pi}{4} \approx 3.9''$$

Mechanical Insight—Test III

Correct Answers

1. B	3. A	5. D	7. C	9. B	11. C	13. A	15. D
2. D	4. C	6. B	8. A	10. D	12. A	14. B	16. C

Explanatory Answers

1. B   Block should move into cutter; blade should cut into block.

2. D   Moving the weight to the left decreases the movement caused by the weight.

3. A   Each revolution of the arm moves the slotted disc 90° to the next slot.

4. C   Each complete turn of the worm will turn the gear one tooth. Hence, gear will turn very slowly.

5. D   Gears 3 and 4 rotate clockwise.

6. B   ¼ turn-extreme left; ½ turn-middle; ¾ turn-extreme right.

7. C   Centripetal force.

8. A   Valve is closed at a cam radius of x. In the proper position it rises a distance y.

9. B   They expand at different rates.

10. D

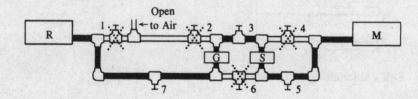

11. C   Twenty turns of the wheel moves the system one inch. Block moves one inch to the right.

12. A $.2 + .025 + .002 = .2270$

13. A   Moving nut C to the left adds clockwise moment (torque) in lieu of nut A.

14. B   Inversely proportional.

     A = RPM of 7″ diameter; B = RPM of Y

$$\underbrace{(10)\,(100)}_{\text{pulley X}} = 7a$$

$$a = \frac{1000 \text{ RPM}}{7}$$

$$14a = \overbrace{\frac{5b}{\text{pulley Y}}}$$

$$14 \left(\frac{1000}{7} \right) = 5b$$

$$b = 400 \text{ RPM}$$

15.   D   Build up from smallest diameter.

16.   C   Stretch of springs end to end, where each spring supports the total weight of 5 pounds, would be twice as much (C) as when the two springs are in parallel and each supports half the load.

Mechanical Insight—Test IV

Correct Answers

1. B	5. B	9. A	13. C	17. C
2. C	6. C	10. D	14. B	18. B
3. C	7. C	11. C	15. D	19. A
4. D	8. C	12. B	16. C	

Explanatory Answers

1.   B   The greater the radius at which the driving force is applied, the slower the rotation.

2.   C   Different speeds due to position of driver from wheel axes. Same directions.

3.   C   Two threads are 20°. One turn is one thread is 10°.

4.   D   Opens valve $\dfrac{\text{once}}{\text{sec.}} = 1 \dfrac{\text{rod}}{\text{sec.}} = \dfrac{10 \text{ rods}}{10 \text{ sec.}} \left(\dfrac{1 \text{ revolution}}{10 \text{ rods}} \right) \left(\dfrac{60 \text{ sec.}}{1 \text{ min.}} \right) = \dfrac{600}{100} = 6 \text{ RPM}$

5.   B   One turn winds a distance equal to the circumference.

6.   C   In one revolution W1 moves down 20, W2 moves up 20. The difference is 40.

7.   C $L = \frac{3}{4} + 1\frac{3}{16} + 2\frac{1}{8} + \frac{5}{32}$

          $= \dfrac{24 + 38 + 68 + 5}{32}$

          $= {}^{135}/_{32}$

          $= 4^{7}/_{32}$ inches

8.   C   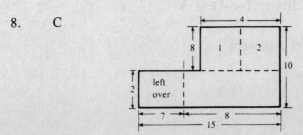   Three 2×8 rectangles.

9.   A

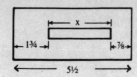

slot = x = 5½ − (1¾ + ⅞)
              = 5⁴/₈ − 2⁵/₈
              = 2⅞ inches

10.   D

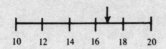

17.

11.   C

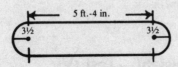

½(2πr) + 2(5 ft.-4 in.) + ½(2πr)
= ½·[2(²²/₇)(⁷/₂)] + 2[5 ft.-4 in.] + ½[2(²²/₇)(⁷/₂)]
= 11 in. + 10 ft.-8 in. + 11 in.
= 10 ft.-30 in.
= 12 ft.-6 in.
= 12 feet, 6 inches

12.   B

2 + 10 + 2 = 14 inches

13.   C   The added water pressure will cause the water to rise, compressing the air.

14.   B   The discharge pipe is above the floor level of the tank, therefore the tank always has a residual amount of water.

15.   D   As the cross-sectional area decreases, the velocity increases. The volume of water flow is constant.

16.   C   Points 1 and 2 travel at the same RPM, but point 2 travels a greater arc length.

17.   C   For a frictionless system in equilibrium, an external force will cause the system to move continuously (until it hits a solid object such as the floor).

18.   B   Part 2 is fastened directly. See diagram.

19.   A   The supporting ropes each carry equal loads.

Mechanical Insight—Test V

Correct Answers

1. D	5. D	9. A	13. C	17. C
2. A	6. C	10. B	14. C	18. A
3. A	7. A	11. B	15. A	19. A
4. A	8. A	12. C	16. B	20. B

Explanatory Answers

1.   D   The supporting edge of the brace under the shelf in fig. D is at the extreme forward edge, thus there is no length of arm for a downward force to produce a torque.

2.   A

d_1 d_2 $f_1 d_1 = f_2 d_2$. Since $d_2 > d_1$, $f_1 > f_2$.

f_1 f_2

75

3.   A   Points Q-R are at the greatest distance from the center of the cam, hence Q-R will raise the follower to its highest position.

4.   A   The shorter wire has a shorter wavelength and a higher frequency.

5.   D   Gaining speed in descent causes apparent weight loss, in ascent causes apparent weight gain.

6.   C   At its highest point the speed of the ball is 0.

7.   A   Inversely proportional.
           (teeth) (rev.) = (teeth) (rev.)
               (15) (14) = (10) (x)
                     x = 21 revolutions

8.   A   A turns faster than B; B speed = C speed; C turns faster than D. A is the fastest.

9.   A   Once, when the cam reached a larger radius.

10.   B   Adjacent gears turn in opposite directions. Gear A and gear B both turn in the same direction.

11.   B   If shaft x turns in direction 2, gear A will turn in the same direction, and mating bevel gear on shaft F will turn in direction 3.

12.   C   150(12) = 600x
               x = 3 ft. in the opposite direction.

13.   C   Brace C, which extends to the ends of the angles, provides the least leverage if stress is applied to bend one of the sides.

14.   C   Gear B can slide on its shaft and only mesh with gear E, which is of suitable diameter.

15.   A   Atmospheric pressure decreases with altitude.

16.   B   Gear C driven by gear on shaft x will turn in direction 3. Gear E will turn in direction 4.

17.   C   The siphon works because of air pressure on the top of the liquid.

18.   A   As the velocity increases (under tube B), the pressure decreases.

19.   A   The same RPM, different direction of rotation.

20.   B   The same RPM, different direction of rotation.

MECHANICAL KNOWLEDGE TESTS

On the theory that mechanical aptitude goes hand in hand with mechanical interest, many mechanical aptitude tests draw upon specific mechanical knowledge. In this chapter, we have assembled a wide variety of actual test questions from civil service and military examinations that are used to measure mechanical knowledge. Answer as many of these questions as possible. The principles on which they are based may well appear on the test you have to take.

TEST I. SHOP PRACTICES

DIRECTIONS: For each question, select the choice which best answers the question or completes the statement. Answers for these questions are provided at the end of the chapter.

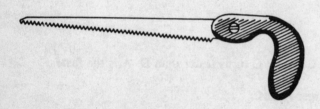

1. The saw shown above is used mainly to cut
   (A) across the grain of wood
   (B) along the grain of wood
   (C) plywood
   (D) odd-shaped holes in wood

2. Concrete is usually made by mixing
   (A) only sand and water
   (B) only cement and water
   (C) lye, cement, and water
   (D) rock, sand, cement, and water

3. The set of a saw is the
   (A) angle at which the handle is set
   (B) amount of springiness of the blade
   (C) amount of sharpness of the teeth
   (D) distance the points stick out beyond the sides of the blade

4. The principal reason for "tempering" or "drawing" steel is to
   (A) reduce stength
   (B) reduce hardness
   (C) increase strength
   (D) increase maleability

5. Sheet metal is dipped in sulphuric acid to
   (A) clean it
   (B) soften it
   (C) harden it
   (D) prevent it from rusting

6. The cut of a file refers to the
   (A) shape of its handle
   (B) shape of its edge
   (C) kind of metal it is made of
   (D) kind of teeth it has

7. In grinding a good point on a twist drill, it is necessary that
   (A) the point be extremely sharp
   (B) both cutting edges have the same lip
   (C) a file be used for the entire cutting process
   (D) the final grinding be done by hand

8. The tool used to locate a point directly below a ceiling hook is a
   (A) a plumb bob
   (B) line level
   (C) transit
   (D) drop gauge

9. The sawing of a piece of wood at a particular angle, for example 45 degrees, is accomplished by using a
   (A) jointer
   (B) cant board
   (C) miter box
   (D) binder

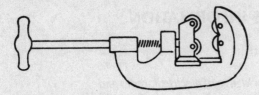

10. The tool above is a
    (A) marking gauge
    (B) knurling tool
    (C) thread cutter
    (D) pipe cutter

11. A high speed grinder operator will check the abrasive wheel before starting the machine because
    (A) it must be wetted properly before use
    (B) if cracked or chipped, it could injure someone
    (C) a dry wheel will produce excessive sparks
    (D) previous work may have clogged the wheel

12. When marking wood, an allowance of 1/16″ to 1/8″ should be made to allow for
    (A) drying of the wood
    (B) absorption of water by wood
    (C) the width of the saw
    (D) knots in the wood

13. A "pinch bar" is used for
    (A) joining
    (B) leveling
    (C) prying
    (D) tightening

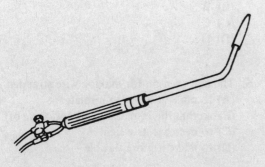

14. The tool shown above is used for
    (A) pressure lubricating
    (B) welding steel plate
    (C) drilling small holes in tight places
    (D) holding small parts for heat treating

15. The primary function of a power driven sabresaw is to
    (A) cut angles
    (B) saw heavy wood stock
    (C) cut curves in flat wood
    (D) make perfectly straight cuts

16. What tool is shown above?
    (A) countersink
    (B) keyhole saw
    (C) hole saw
    (D) grinding wheel

17. The tip of a soldering iron is usually made of
    (A) iron
    (B) steel
    (C) lead
    (D) copper

18. Which of the following is used with a miter box?
    (A) back saw
    (B) keyhole saw
    (C) coping saw
    (D) compass saw

19. The length of a six penny nail is about
    (A) 1 inch
    (B) 2 inches
    (C) 3 inches
    (D) 4 inches

20. High oil content or so-called "spar" varnish is used primarily for
    (A) finishing furniture
    (B) obtaining a high gloss finish
    (C) finishing weather exposed surfaces
    (D) finishing interior trim

TEST II. ELECTRONICS INFORMATION

DIRECTIONS: This is a test of your knowledge of electrical, radio and electronics information. You are to select the correct response from the choices given. Then mark the space on your answer sheet that corresponds to the number and letter of your choice. Answers to these questions will be found at the end of the chapter.

1. The most likely cause of a burned-out fuse in the primary circuit of a transformer in a rectifier is
   (A) grounding of the electrostatic shield
   (B) an open circuit in a bleeder resistor
   (C) an open circuit in the secondary winding
   (D) a short-circuited filter capacitor

2. The primary coil of a power transformer has 100 turns and the secondary coil has 50 turns. The voltage across the secondary will be
   (A) four times that of the primary
   (B) twice that of the primary
   (C) half that of the primary
   (D) one-fourth that of the primary

3. The best electrical connection between two wires is obtained when
   (A) the insulations are melted together
   (B) all insulation is removed and the wires bound together with friction tape
   (C) both are wound on a common binding post
   (D) they are soldered together

4. Excessive resistance in the primary circuit will lessen the output of the ignition coil and cause the
   (A) battery to short out and the generator to run down
   (B) battery to short out and the plugs to wear out prematurely
   (C) generator to run down and the timing mechanism to slow down
   (D) engine to perform poorly and hard to start

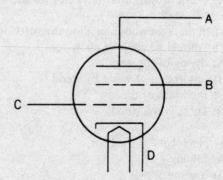

5. In the schematic vacuum tube illustrated, the cathode is element
   (A) A
   (B) B
   (C) C
   (D) D

6. The main reason for making wire stranded is
   (A) to make it easier to insulate
   (B) so that the insulation will not come off
   (C) to decrease its weight
   (D) to make it more flexible

7. The oscilloscope image shown on page 228 represents
   (A) steady DC
   (B) resistance in a resistor
   (C) AC
   (D) pulsating DC

8. Voltage drop in a circuit is usually due to
   (A) inductance
   (B) capacitance
   (C) resistance
   (D) conductance

9. If an increase in grid voltage no longer produces an increase in plate current, the tube has reached its
   (A) inversion point
   (B) saturation point
   (C) class C operating point
   (D) class A operating point

10. Earphones are generally not used with radio receivers having more than three tubes because
    (A) earphones can handle only alternating current
    (B) the amplification factor makes them unnecessary
    (C) only one person may hear them at a time
    (D) earphones are too delicate for normal use

11. Of the non-metallic elements listed below, which one is the best conductor of electricity?
    (A) Mica
    (B) Carbon
    (C) Formica
    (D) Hard rubber

12. If an electric motor designed for use on AC is plugged into a DC source, what will probably happen?
    (A) Excessive heat will be produced
    (B) It will operate the same as usual
    (C) It will continue to operate, but will not get so warm
    (D) It cannot be predicted what will happen

13. Most electrical problems involving voltage, resistance, and current are solved by applying
    (A) Ohm's Law
    (B) Watt's Law
    (C) Coulomb's Law
    (D) Kirchoff's Voltage and Current Laws

14. If every time a washing machine is started the circuit breaker must be reset, the best solution would be to
    (A) oil the motor in the washer
    (B) replace the circuit breaker
    (C) tape the breaker switch closed
    (D) repair the timing mechanism

15. In most AC-DC radio circuits when one tube filament burns out it will
    (A) cause the others to burn out
    (B) open the circuit and keep the others from operating
    (C) cause the remaining ones to operate at higher current ratings
    (D) cause the line voltage to drop

16. The most stable type of radio oscillating circuit is the
    (A) electron-coupled
    (B) crystal
    (C) heterodyne
    (D) colpitts

17. The ampere is the unit of measurement of
    (A) inductance
    (B) resistance
    (C) voltage
    (D) current

18. Hoping to make his car run faster, a "hot-rodder" decides to try changing the ignition mechanism. He finds all the components in good working order, so he decides to
    (A) use a larger capacitor on the points
    (B) retard the ignition several degrees
    (C) put hotter spark plugs in the engine
    (D) check the ignition timing

19. A mixer, in radio terminology, would function to
    (A) jumble a carrier wave for security transmissions
    (B) couple the stages of two succeeding circuits
    (C) coordinate the triodes in a push-pull power amplifier circuit
    (D) combine the incoming and local oscillator frequencies

20. Flux is used in the process of soldering together two conductors in order to
    (A) provide a luster finish
    (B) prevent oxidation when the connection is heated

(C) maintain the temperature of the soldering iron

(D) prevent the connection from becoming overheated

21. Which of the following devices converts heat energy directly into electrical energy?
(A) A piezoelectric crystal
(B) A photoelectric cell
(C) A steam driven generator
(D) A thermocouple

22. One use of a coaxial cable is to
(A) ground a signal
(B) pass a signal from the set to the antenna of a mobile unit
(C) carry the signal from a ballast tube
(D) carry grid signals in high altitude areas

23. Which of the following has the **least** resistance?
(A) silver
(B) aluminum
(C) copper
(D) iron

24. A rectifier is used to convert
(A) alternating current into direct current
(B) static current into direct current
(C) direct current into alternating current
(D) low frequency current into high frequency current

25. The length of a radio transmitter antenna system is primarily determined by
(A) transmitter power
(B) transmitter frequency
(C) oscillator voltage
(D) distance from receiving antenna

26. Which one of the following may best be compared to electrical voltage?
(A) Tension
(B) Resistance
(C) Flow
(D) Pressure

27. The extent to which a radio receiver converts the signals received into sounds that are undistorted is called
(A) fidelity   (C) selectivity
(B) sensitivity   (D) resonance

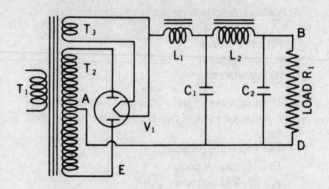

28. The tube in the figure above acts as a
(A) voltage regulator
(B) voltage divider
(C) half-wave rectifier
(D) full-wave rectifier

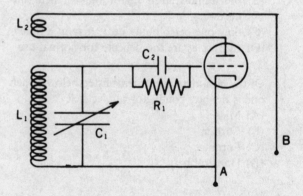

29. The oscillator in the circuit in the figure shown above is known as a
(A) tuned-grid oscillator
(B) tuned-plate oscillator
(C) electron-coupled oscillator
(D) resistance feedback

30. The power supply in a vacuum tube equipped auto radio differs from the power supply in an AC home radio in that the former utilizes a
(A) filter capacitor
(B) choke
(C) vibrator
(D) rectifier tube

TEST III. AUTOMOTIVE INFORMATION

DIRECTIONS: This test has questions about automobiles. Select the best answer for each question, then blacken the space on your answer sheet that corresponds to the number and letter of your answer choice. Check the answer key at the end of the chapter to see how well you did.

1. Which of the following devices prevents the generator / alternator from overcharging the battery in an automobile?
   (A) Governor
   (B) Solenoid switch
   (C) Current regulator
   (D) Voltage regulator

2. A torsion bar might be found in the
   (A) transmission
   (B) distributor
   (C) speedometer
   (D) suspension

3. A black gummy deposit in the end of the tail pipe of an automobile indicates that
   (A) the automobile "burns" oil
   (B) there is probably a leak in the exhaust manifold
   (C) the timing is late
   (D) there are leaks in the exhaust valves

4. What would be the most probable cause if an automobile has a weak spark at the plugs, "turns over" very slowly, and has dim headlights?
   (A) Weak battery
   (B) Faulty condenser
   (C) Faulty ignition cable
   (D) Worn contact breaker points

5. An automobile engine won't "turn over." If the battery charge is found to be normal, the next test would normally be for
   (A) defective starter motor
   (B) short-circuited switches
   (C) faulty battery cable connections
   (D) defective generator

6. The generator or alternator of an automobile engine is usually driven by the
   (A) camshaft
   (B) flywheel
   (C) fan belt
   (D) cranking motor

7. What source of trouble can be tested by removing a spark plug and holding a thumb over the spark plug hole while the engine is being cranked?
   (A) Poor ignition
   (B) Low compression
   (C) High oil consumption
   (D) High fuel consumption

8. If an automobile engine overheats while the radiator remains cold, the difficulty probably lies in
   (A) lack of engine oil
   (B) stuck thermostat
   (C) improper ignition timing
   (D) an overloaded engine

9. It is best for an automobile's gas tank to be full or nearly-full to prevent
   (A) gasoline from vaporizing in the fuel lines
   (B) moisture from condensing in the gas tank
   (C) drying out of the fuel pump
   (D) loss of vacuum in the vacuum line

10. In troubleshooting the rear axle, an automobile is driven on a smooth road at 25 mph, and the accelerator is lightly pressed and released. If there is a "slapping" noise, the most probable trouble is
    (A) a worn universal joint
    (B) an incorrect drive line angle
    (C) loose accelerator linkage
    (D) a bent transmission shaft

11. An automobile handbrake is set tightly and the engine is idling at 30 mph road speed. If you shift into high gear, release the clutch, and the engine continues to run about the same, what would most likely need repair?
    (A) Clutch
    (B) Throttle
    (C) High gear
    (D) Carburetor

12. The pistons of gasoline engines will sometimes increase in size so that they "stick" in the cylinder. This is often caused by
    (A) low engine operating temperature
    (B) overheating of the engine
    (C) worn oil rings
    (D) worn compression rings

13. If a 4-speed transmission makes noise when engaged in low gear, it will likely also make noise when in
    (A) 3rd gear
    (B) 4th gear
    (C) reverse
    (D) neutral

14. The letter "W" in the oil designation SAE 20-W means
    (A) the oil was produced by a refinery in a western state
    (B) the oil is adapted for cold weather starting
    (C) the oil is water-soluble
    (D) it is a flushing oil

15. What will happen if leaded gasoline is used in a car equipped with a catalytic converter
    (A) intake valves will crack
    (B) catalytic converter will be damaged
    (C) engine will overheat
    (D) engine will begin backfiring

16. Dual or multiple carburetors are used to obtain
    (A) a richer mixture of air and gasoline
    (B) a more uniform distribution of fuel charge
    (C) an overlapping of suction periods on one mixing tube
    (D) flexibility of firing order

17. If an automobile air conditioning system fails to cool, the first check to make is for
    (A) leaks in hoses
    (B) malfunction of the compressor
    (C) low oil level
    (D) shortage of refrigerant

18. If the air cleaner on an automobile engine becomes clogged, the effect on engine performance will be similar to that of a
    (A) partly closed choke valve
    (B) vapor lock
    (C) clogged fuel nozzle
    (D) needle valve stuck in closed position

19. Which of the following instruments can be used to adjust the air fuel ratio, valve timing and check for leaky head gaskets?
    (A) Compression tester
    (B) Vacuum gauge
    (C) Timing light
    (D) Dwell meter

20. In the operation of a gasoline engine, ignition coil failure is most often caused by
    (A) a low battery
    (B) an overcharged battery
    (C) burned coil terminal
    (D) moisture entering coil case

TEST IV. MAINTENANCE WORK

DIRECTIONS: Maintenance work depends heavily on mechanical aptitude and knowledge. For this reason we have assembled a variety of actual test questions for exams given to maintenance workers. Try this test to see how your own mechanical knowledge measures up. Answers are provided at the end of the chapter.

1. The practice of placing extra weight on the rear of a fork-lift truck which is carrying an overload is
   (A) undesirable, because the operator has too much balancing to do
   (B) undesirable, because it puts a strain on the motor, tires, and axle of the truck
   (C) desirable, because this prevents the truck from turning over
   (D) desirable, because more material can be transported at a time

2. Of the following, the most important reason for not letting oily rags accumulate in an open storage bin is that they
   (A) may start a fire by spontaneous combustion
   (B) will drip oil onto other items in the bin
   (C) may cause a foul odor
   (D) will make the area messy

3. Of the following, the best method to employ in putting out a gasoline fire is to
   (A) use a bucket of water
   (B) smother it with rags
   (C) use a carbon dioxide extinguisher
   (D) use a carbon tetrachloride extinguisher

4. Assume that you have to move ten 65-pound crates a distance of approximately 350 feet and each crate measures 14″ x 26″ x 32″. From among the following methods, it would be best to
   (A) load the crates on a pallet and use a fork-lift truck
   (B) carry one crate at a time by yourself
   (C) load the crates on a skid and use pipe rollers to move the skid
   (D) unpack each crate and move all of the contents with a motor van

5. Of the following, the best reason for stacking long rectangular tubes in layers, with the first layer lengthwise and the next layer crosswise, is that it
   (A) reduces the overall stacking height
   (B) makes it simpler to count the tubes
   (C) makes it easier to remove a tube from the center of the stack
   (D) prevents the stack of tubes from toppling

6. The accepted practice for a person to follow in lifting a heavy object off the floor is to
   (A) keep both legs straight and close together, and to bend at the waist to grasp the object
   (B) get a solid footing, and with both legs straight, bend at the waist and lift the object
   (C) place the feet as far apart as possible and bend at the knees to reach down to grasp the object
   (D) place the feet shoulder width apart and bend at the knees to reach down to grasp the object

7. When a new shipment of material is received, it is sometimes necessary to store the new material in such a way that the old stock will be used first. It is most important to use this method with material that
   (A) is ordered in large quantities
   (B) is large in size
   (C) is not used often
   (D) deteriorates with age

8. Of the following, the best reason for storing small items, such as nails, in their original containers whenever possible is that it
   (A) makes it easier to inspect these items
   (B) eliminates the need for bins and shelves
   (C) makes it simpler to identify these items
   (D) reduces the loss of the item due to theft.

9. Of the following, the one that is a grease fitting is a
   - (A) brown fitting
   - (B) zerk fitting
   - (C) taper fitting
   - (D) morse fitting

10. Of the following, the best tool to use to make a hole in a concrete floor for a machine hold-down bolt is a
    - (A) counterboring tool
    - (B) cold chisel
    - (C) drift punch
    - (D) star drill

11. Of the following, the best type of saw to use to cut a 4-inch diameter hole through a 5/8-inch wooden partition is a
    - (A) back saw
    - (B) saber saw
    - (C) circular saw
    - (D) cross-cut saw

12. When removing a shrink-fitted collar from a shaft, it would be easiest to drive out the shaft after
    - (A) heating only the collar
    - (B) heating only the shaft
    - (C) chilling only collar
    - (D) chilling the collar and heating the shaft

13. Of the following, the best reason for overhauling a machine on a regular basis is
    - (A) that overhauling is easier to do when done often
    - (B) to minimize breakdowns of the machine
    - (C) to make sure the machine is properly lubricated
    - (D) to make sure the employees are familiar with the machine

14. While using a hacksaw to cut through a one-inch diameter steel bar, one should not press down too heavily on the hacksaw because this may
    - (A) break the blade
    - (B) overheat the bar
    - (C) permanently distort the frame
    - (D) cause the hacksaw to slip

15. A miter box is used
    - (A) for locating dowel holes in two pieces of wood to be joined together

(B) to hold a saw at a fixed angle while sawing
(C) to hold a saw while sharpening its teeth
(D) to clamp two pieces of wood together at 90 degrees

16. Wing nuts are especially useful on equipment where
    - (A) the nuts must be removed frequently and easily
    - (B) the nuts are locked in place with a cotter pin
    - (C) critical adjustments are to be made frequently
    - (D) a standard hex head wrench cannot be used

17. The best device to employ to make certain that two points, separated by an unobstructed vertical distance of 12 feet, are in the best possible vertical alignment is a
    - (A) carpenter's square
    - (B) level
    - (C) folding ruler
    - (D) plumb bob

18. In a shop, snips should be used to
    - (A) hold small parts steady while machining them
    - (B) cut threaded pipe
    - (C) cut thin gauge sheet metal
    - (D) remove nuts that are seized on a bolt

19. Caulking a joint means
    - (A) applying sealing material to the joint
    - (B) tightening the joint with wrenches
    - (C) opening it with wrenches
    - (D) testing the joint for leaks

20. When storing files, the most important reason for making sure that the files do not touch each other is to prevent
    - (A) damage to the file teeth
    - (B) damage to the file shanks
    - (C) rusting of the files
    - (D) dirt from accumulating in the file teeth

21. A clutch is a device that is used
    - (A) to hold a work piece in a fixture
    - (B) for retrieving small parts from hard to reach areas
    - (C) to disengage one rotating shaft from another
    - (D) to level machinery on a floor

22. Of the following, the best device to use to determine whether the surface of a work bench is horizontal is a
    (A) surface gage
    (B) spirit level
    (C) dial vernier
    (D) profilometer

23. Of the following, the machine screw having the smallest diameter is the
    (A) 10-24 x 3/4″
    (B) 6-32 x 1 1/4″
    (C) 12-24 x 1″
    (D) 8-32 x 1 1/2″

24. When drilling into a steel plate, the most likely cause for the breaking of a drill bit is
    (A) too low a drill speed
    (B) excessive cutting oil lubricant
    (C) too much drill pressure
    (D) using a bit with a dull point

25. Of the following, the most important advantage of a ratchet wrench over an open-end wrench is that the ratchet wrench
    (A) can be used in a more limited space
    (B) measures the torque applied
    (C) will not strip the threads of a bolt
    (D) is available for all sizes of hex bolts

26. The nominal voltage of the "D" size dry-cell battery used in common hand-held flashlights is most nearly
    (A) 1 volt
    (B) 1.5 volts
    (C) 2.0 volts
    (D) 2.5 volts

27. In an electric circuit, a volt-ohmmeter can be used to directly measure
    (A) inductance
    (B) power
    (C) resistance
    (D) capacitance

28. An ammeter is a device used for measuring the
    (A) current in an electrical circuit
    (B) dimensions of small mechanical parts
    (C) voltage in an electric circuit
    (D) depth of holes

29. The purpose of a water trap in a plumbing drainage system is to
    (A) prevent the leakage of water
    (B) prevent freezing of the pipes
    (C) block off sewer gases
    (D) reduce the water pressure in the system

30. Small leaks in a compressed air pipe line leading from a shop compressor are most easily located by
    (A) creating a vacuum in the air line
    (B) allowing the compressor to pump water through the lines
    (C) monitoring air gauges throughout the piping system
    (D) applying soapy water to the pipeline

31. The tool that holds the die when threading pipe is generally called a
    (A) vise
    (B) stock
    (C) yoke
    (D) coupling

32. A fitting used to join a small pipe at right angles to the middle of a large pipe is called a
    (A) union
    (B) coupling
    (C) cap
    (D) reducing tee

33. Gaskets are commonly used between the flanges of large pipe joints to
    (A) make a leakproof connection
    (B) provide for expansion
    (C) provide space for assembly
    (D) adjust for poor alignment

34. The pipe fitting that should be used to connect a 1″ pipe to a 1 1/2″ valve is called a
    (A) reducing coupling
    (B) nipple
    (C) bushing
    (D) union

35. To prevent damage to an air compressor, the air coming into the compressor is usually
    (A) cooled
    (B) heated
    (C) expanded
    (D) filtered

36. The reason for galvanizing sheet metal is to
    (A) make it harder
    (B) increase its tensile strength
    (C) prevent it from being a conductor of electricity
    (D) make it rust-resistant

37. A hole drilled in a shaft would probably be reamed to fit a
    (A) lag screw
    (B) cap screw
    (C) carriage bolt
    (D) taper pin

38. The part of a drill press which is used to hold the drill bit is called a
    (A) chuck
    (B) collar
    (C) bit
    (D) vise

39. The part of a bus that allows one rear wheel to turn faster or slower than the other when turning a corner, is the
    (A) universal joint
    (B) rear axle
    (C) idler
    (D) differential

40. In a 4-stroke cycle diesel engine, the fuel is ignited by means of
    (A) compressed air at a high temperature
    (B) special spark plugs
    (C) cold spark plugs
    (D) hot spark plugs

41. The basic purpose of an idler gear in a gear train is to
    (A) change gear speed
    (B) increase gear torque
    (C) reduce friction in the gear train
    (D) change the direction of rotation of a shaft

42. When cutting a left-hand thread on a lathe, it is necessary to reverse the direction of the
    (A) chuck
    (B) driving motor
    (C) lead screw
    (D) lathe centers

43. In order to cut a 2 inch diameter hole accurately into a sheet of 16 gauge sheet metal, it is best to use a
    (A) cutter and a bar
    (B) hand reamer
    (C) high speed drill
    (D) nibbler

44. The instrument that is commonly used to check the armature of small d.c. motors for shorts, grounds or an open circuit is
    (A) an ammeter
    (B) a dynamometer
    (C) a growler
    (D) a voltmeter

45. In some plant operations, d.c. current is required where only a.c. is supplied by Con Edison. A device that is used to convert the a.c. to d.c. current is called
    (A) an inductor coil
    (B) a motor-generator
    (C) a rheostat
    (D) a transformer

46. Of the following, the most important use for a flexible coupling is to connect two shafts which may
    (A) rotate in opposite directions
    (B) have different diameters
    (C) occasionally become slightly misaligned
    (D) rotate at different speeds

47. The purpose of the packing which is generally found in the stuffing box of a centrifugal pump is to
    (A) prevent the impeller from chattering
    (B) prevent the leakage of fluid
    (C) reduce bearing wear
    (D) reduce the discharge pressure

48. A bus wheel which is unbalanced, should be rebalanced by
    (A) retreading the tire
    (B) bending the rim slightly
    (C) replacing the wheel bearing
    (D) adding weights at the rim

49. "Truing" a grinding wheel refers to
    (A) making the face of the wheel parallel to the spindle
    (B) centering the wheel mounting hole
    (C) making the face of the wheel larger
    (D) mounting the wheel onto the spindle

50. A flux is applied during a brazing operation primarily to
    (A) prevent fusion and penetration throughout the joint
    (B) prevent formation of oxide films in the area of the joint
    (C) reduce the electrical conductivity of the joint
    (D) reduce the surface hardness in the area of the joint.

51. When grinding a flat chisel, it is good practice to keep the chisel moving across the face of the grinding wheel in order to prevent
    (A) grooving of the wheel
    (B) burning of the chisel tip
    (C) the wheel from vibrating
    (D) the wheel from cracking

52. An electrical ballast is used in
    (A) a heavy duty electric power drill
    (B) a motor-generator set
    (C) an electrical circuit breaker
    (D) a fluorescent lighting system

53. An electrical transformer can be used to
    (A) raise battery output voltage
    (B) maintain constant battery output voltage
    (C) lower the voltage from a 110 volt a.c. power line
    (D) change the current from a.c. to d.c.

54. Metals are commonly arc welded electrically by the use of
    (A) high voltage and high current
    (B) high voltage and low current
    (C) low voltage and high current
    (D) low voltage and low current

55. Before drilling a hole in a steel plate, an indentation should be made with a
    (A) center punch
    (B) nail
    (C) drill bit
    (D) pin punch

Correct Answers
For Mechanical Knowledge Tests

SHOP PRACTICES TEST I

1. D	5. A	9. C	13. C	17. D
2. D	6. D	10. D	14. B	18. A
3. D	7. B	11. B	15. C	19. B
4. C	8. A	12. C	16. C	20. C

ELECTRONICS INFORMATION TEST II

1. D	6. D	11. B	16. B	21. D	26. D
2. C	7. D	12. A	17. D	22. B	27. A
3. D	8. C	13. A	18. D	23. A	28. D
4. D	9. B	14. B	19. D	24. A	29. A
5. D	10. B	15. B	20. B	25. B	30. C

AUTOMOTIVE INFORMATION TEST III

1. D	5. C	9. B	13. C	17. D
2. D	6. C	10. A	14. B	18. A
3. A	7. B	11. A	15. B	19. B
4. A	8. B	12. B	16. B	20. D

MAINTENANCE WORK TEST IV

1. B	12. A	23. B	34. C	45. B
2. A	13. B	24. C	35. D	46. C
3. C	14. A	25. A	36. D	47. B
4. A	15. B	26. B	37. D	48. D
5. D	16. A	27. C	38. A	49. A
6. D	17. D	28. A	39. D	50. B
7. D	18. C	29. C	40. A	51. A
8. C	19. A	30. D	41. D	52. D
9. B	20. A	31. B	42. C	53. C
10. D	21. C	32. D	43. A	54. C
11. B	22. B	33. A	44. C	55. A

All the answers to the mechanical knowledge tests
are self-explanatory.

SHOP ARITHMETIC
AND CALCULATION

Most mechanical work requires facility with the basic operations of arithmetic necessary to make calculations on the job. To give you an idea of the kind of numerical ability expected, we have compiled this selection of sample questions from tests for a variety of mechanical jobs. Difficulty in solving these problems, indicates a need to brush-up on your computational and problem-solving skills. Answers to all Shop Arithmetic Tests are provided at the end of the chapter.

(Pi $= \pi \approx \frac{22}{7}$, $3\frac{1}{7}$ or 3.14)

TEST I. SHOP ARITHMETIC

DIRECTIONS: For each question in this test, read carefully the stem and the four lettered choices that follow. Choose the answer which you consider correct or most nearly correct. Mark the answer sheet for the letter you have chosen: A, B, C, or D. Check your answers with the Correct Answers at the end of the chapter.

1. A certain type of paint is capable of covering about 400 square feet of wall surface per gallon. How many gallons of this type of paint will be required to cover a wall that measures 73 feet by 15 feet?
   (A) 1 1/2 gallons
   (B) 2 3/4 gallons
   (C) 5 gallons
   (D) 6 1/2 gallons

2. A train leaves Terminal "a" for Terminal "b;" at the same time, another train leaves Terminal "b" for Terminal "a." Train leaving "a" travels at the rate of 45 miles per hour. Train leaving "b" travels at the rate of 25 miles per hour. The distance between "a" and "b" is 1540 miles. How long after leaving Terminals will the two trains pass?
   (A) 20 hours
   (B) 21 hours
   (C) 21 1/2 hours
   (D) 22 hours

3. A steam shovel excavates 2 cubic yards every 40 seconds. At this rate, the amount excavated in 45 minutes is
   (A) 90 yds.³
   (B) 135 yds.³
   (C) 900 yds.³
   (D) 3,600 yds.³

Questions 4 to 7 are based on the information given in the following paragraph.

The four departments of a railroad shop occupy floor space as follows:

Inspector shop, 3300 square feet; machine shop, 1700 square feet; paint shop, 1500 square feet; truck shop, 2000 square feet. The yearly heating cost for the entire shop is $1870. This expense is distributed on the basis of the floor space occupied by each department.

4. What amount should be charged to the inspection shop?
   (A) $726
   (B) $374
   (C) $440
   (D) $330

5. What amount should be charged to the machine shop?
   (A) $726
   (B) $374
   (C) $440
   (D) $330

253

6. What amount should be charged to the paint shop?
   (A) $726
   (B) $374
   (C) $440
   (D) $330

7. What amount should be charged to the truck shop?
   (A) $726
   (B) $374
   (C) $440
   (D) $330

8. If measured accurately, 36 square yards of floor surface will be found to contain
   (A) 324 ft.²
   (B) 800 ft.²
   (C) 200 ft.²
   (D) 144 ft.²

9. The hypotenuse of a right triangle whose sides are 3 feet and 4 feet is
   (A) 7 feet
   (B) 9 feet
   (C) 5 feet
   (D) 8 feet

10. One brass rod measures 3 5/16 inches long and another brass rod measures 2 3/4 inches long. Together their length is
    (A) 6 9/16 inches
    (B) 5 1/8 inches
    (C) 6 1/16 inches
    (D) 5 1/16 inches

11. 4000 sheets of 8" x 11" paper are purchased at $12.00 a ream for 32" x 44" stock. If there are 500 sheets in a ream the cost is
    (A) $12.00
    (B) $24.00
    (C) $ 6.00
    (D) $ 5.00

12. 20 board feet of lumber are purchased at $120 per C. The cost is
    (A) $12.00
    (B) 60¢
    (C) $2.40
    (D) $24.00

13. The sum of 5 feet 2 3/4 inches; 8 feet 1/2 inch; and 12 1/2 inches is
    (A) 14 feet, 3 3/4 inches
    (B) 14 feet, 5 3/4 inches
    (C) 14 feet, 9 1/4 inches
    (D) 15 feet, 1/2 inch

14. A train traveling at 30 miles per hour will go one mile in
    (A) 1/2 minute
    (B) 2 minutes
    (C) 4 minutes
    (D) 5 minutes

15. The number of sacks of cement necessary for a 4 inch sidewalk 6 ft. wide and 27 ft. long at 6 sacks of cement per cubic yard of concrete is
    (A) 12
    (B) 6
    (C) 36
    (D) 9

16. A barrel, containing an equal number of bolts and nuts, weighs 60 lbs. If each bolt weighs 1/6 of a lb. and is 5 times as heavy as the nut, the number of nuts in the barrel is
    (A) 300
    (B) 360
    (C) 50
    (D) 72

17 If iron weighs 0.25 lb. per cubic inch, an iron bar 8'-6" by 4" by 1/2" weighs
    (A) 4.25 lbs.
    (B) 51 lbs.
    (C) 68 lbs.
    (D) 204 lbs.

18. A mechanic who receives $9.75 per hour, and works 8 hours a day for 5 days, will earn a total of
    (A) $350.00
    (B) $360.00
    (C) $390.00
    (D) $420.00

19. The circumference of a circle is given by the formula $C = 2 \pi R$, where C is the circumference, R is the Radius, and π is approximately $\frac{22}{7}$ or $3\frac{1}{7}$.

The circumference of an oil drum having a diameter of 1′9″ is most nearly
(A) 132″
(B) 66″
(C) 33″
(D) 17″

20. The area of a circle having a diameter of one inch is closest to
(A) 3/4 in.²
(B) 1 in.²
(C) 1 1/3 in.²
(D) 1 1/2 in.²

21. The total surface area of a 5-inch solid cube is
(A) 100 in.²
(B) 125 in.²
(C) 150 in.²
(D) 200 in.²

22. To change cubic feet into cubic yards
(A) multiply by 27
(B) multiply by 3
(C) divide by 3
(D) divide by 27

23. Using measuring cans without any intermediate marks, two gallons of oil can be accurately measured from a barrel and put in a bearing using
(A) an 8 gallon and a 4 gallon can
(B) two 4 gallon cans
(C) a 6 gallon and a 4 gallon can
(D) a 1 1/2 gallon and a 6 gallon can

24. Six men can wire 3 rooms in 4 hours. At this rate, 4 men can wire 2 rooms in
(A) 2 hours
(B) 4 hours
(C) 4 1/2 hours
(D) 6 hours

25. The number of 125-pound weights that can be lifted safely with a chain hoist of half-ton capacity is
(A) 4
(B) 8
(C) 10
(D) 14

26. A coil of wire consisting of 30 turns of wire has an average diameter of 14 inches. The approximate total length of wire is
(A) 110 feet
(B) 225 feet
(C) 420 feet
(D) 1320 feet

27. If No. 9 wire weights twice as much per foot as No. 12, and No. 9 wire weighs 40 lbs. per 1,000 feet, then 500 feet of No. 12 will weigh
(A) 80 lbs.
(B) 40 lbs.
(C) 20 lbs.
(D) 10 lbs.

28. If copper weighs 0.32 pounds per cubic inch, six feet of copper bus bar 1/4 inch thick and 1 inch wide will weigh
(A) 0.48 pounds
(B) 1.92 pounds
(C) 5.76 pounds
(D) 7.68 pounds

29. An agent travels a total of 4.1 miles each day to and from work. The traveling consumes 73 minutes each day. Approximately how many hours would he save in 129 working days if he moved to another residence so that he would travel only 1.7 miles each day in travel (assuming the same rate of travel)?
(A) 78 hours
(B) 87 hours
(C) 88 hours
(D) 92 hours

30. A shipment consists of 340 ten-foot pieces of conduit with a coupling on each piece. If the conduit weighs 0.85 lbs. per foot and each coupling weighs 0.15 lbs., the total weight of the shipment is
(A) 340 lbs.
(B) 628 lbs.
(C) 2941 lbs.
(D) 3400 lbs.

31. The diameter of a certain rod is required to be 1.51″, ± .015″. The rod would *not* be acceptable if the diameter was
(A) 1.490″
(B(1.500″
(C) 1.510″
(D) 1.525″

32. If a plant making bricks turns out 1,250 bricks in 5 days, the number of bricks which can be made in 20 days is
    (A) 5,000
    (B) 6,250
    (C) 12,500
    (D) 25,000

33. Comparing the cost of a 25-watt lamp burning for 100 hours and the cost of a 100-watt lamp burning for 25 hours, the cost will be

    (A) four times as much for the 100-watt lamp
    (B) the same
    (C) four times as much for the 25-watt lamp
    (D) eight times as much for the 100-watt lamp

34. Seven thirty-seconds of an inch is most nearly
    (A) .007 inches
    (B) .022 inches
    (C) .219 inches
    (D) .319 inches

TEST II. SHOP ARITHMETIC

DIRECTIONS: *For each question in this test, read carefully the stem and the four lettered choices that follow. Choose the answer which you consider correct or most nearly correct. Mark the answer sheet for the letter you have chosen: A, B, C, or D.*

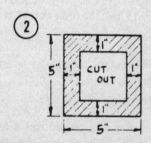

1. The maximum number of gaskets shown which can be cut from the gasket material as shown is
    (A) 14
    (B) 15
    (C) 18
    (D) 20

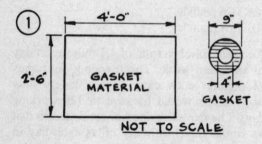

2. The gasket shown has an area of
    (A) 9 in.²
    (B) 15 in.²
    (C) 16 in.²
    (D) 20 in.²

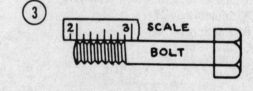

3. The number of threads per inch on the bolt is
    (A) 16
    (B) 10
    (C) 8
    (D) 7

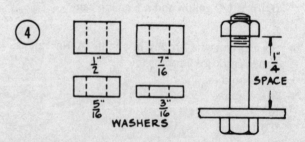

4. Using only the sizes of washers shown, the least number of washers needed to exactly fill the 1 1/4" space is
    (A) 6
    (B) 5
    (C) 4
    (D) 3

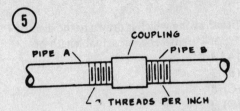

5. If both pipes A and B are free to move back and forth but are held so they cannot turn, and the coupling is turned 4 revolutions with a wrench, the overall length of the pipes and coupling will
   (A) decrease 1/2″
   (B) remain the same
   (C) increase or decrease 1″ depending upon the direction of turning
   (D) increase 1/2″

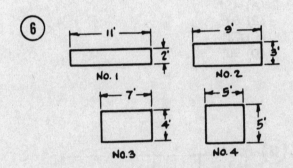

6. Shown are the bottoms of four bins for storing materials. If the bins are all capable of holding the same amount of any particular material, then you would expect the bin with the least height of sides to be the one whose bottom is shown as
   (A) No. 1
   (B) No. 2
   (C) No. 3
   (D) No. 4

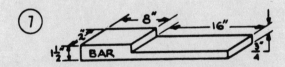

7. The volume (in cubic inches) of the bar is
   (A) 24 in.³
   (B) 28 1/4 in.³
   (C) 48 in.³
   (D) 60 in.³

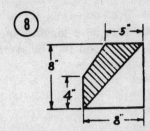

8. If the shaded portion is cut from the plate shown, the area of the remaining portion is
   (A) 26 in.²
   (B) 29 in.²
   (C) 32 in.²
   (D) 58 in²

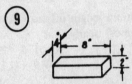

9. The approximate dimensions of a common brick are as shown. The volume of the brick is
   (A) 64 ft.³
   (B) 5 1/3 ft.³
   (C) 4/9 ft.³
   (D) 1/27 ft.³

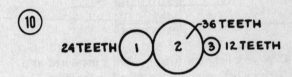

10. When the RPM of gear #1 is 120, the RPM of #3 is
    (A) 40
    (B) 60
    (C) 180
    (D) 240

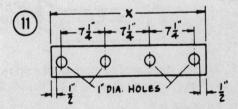

11. The dimension "X" on the piece shown is
    (A) 20 3/4″   (C) 23 3/4″
    (B) 22 3/4″   (D) 24 1/4″

12. Five metal bars have lengths, measured in feet, of 4, 4, 3, 4 and 10. The average (arithmetic mean) length is
    (A) 6.7 feet
    (B) 5 feet
    (C) 4 feet
    (D) 3.8 feet

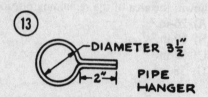

13. The minimum length of strap iron to make the hanger is most nearly
    (A) 26″   (C) 13″
    (B) 15″   (D) 9 1/2″

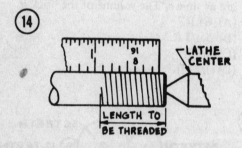

14. A very light cut (trace) is being measured as a check before cutting the thread on the lathe. The number of threads per inch shown is
    (A) 12   (C) 14
    (B) 13   (D) 15

15. The minimum area of sheet metal strip required to make the duct (allowing 2 1/2% for joints) is

    (A) 24.6 ft.²
    (B) 41 ft.²
    (C) 42.5 ft.²
    (D) 100 ft.²

16. In a drawing which is drawn to the scale of 1″ = 1′, 3/4″ represents an actual length of
    (A) 3/4″
    (B) 3″
    (C) 8″
    (D) 9″

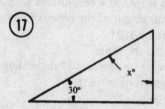

17. In the right-angled triangle shown, x is
    (A) 45°
    (B) 60°
    (C) 78°
    (D) 90°

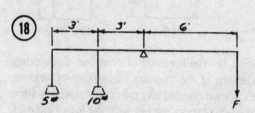

18. The force F needed to balance the lever is most nearly
    (A) 7.5 lbs.
    (B) 10 lbs.
    (C) 12.5 lbs.
    (D) 15 lbs.

19. A drawing uses a scale of 1 inch to represent 10 feet. If a square on the drawing measures 5 in. on a side, what is the actual area?
    (A) 50 ft.²   (C) 2500 in.²
    (B) 2500 ft.²   (D) .25 ft.²

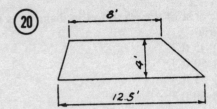

20. The top and bottom sides of the figure shown are parallel. The area is most nearly
    (A) 40.5 ft.²
    (B) 41.0 ft.²
    (C) 41.5 ft.²
    (D) 42.0 ft.²

21. In the circle shown, the radius is 10′. The area of the shaded portion is most nearly
    (A) 27.5 ft.²
    (B) 28.0 ft.²
    (C) 28.5 ft.²
    (D) 29.0 ft.²

TEST III. SHOP ARITHMETIC

DIRECTIONS: For each question in this test, read carefully the stem and the four lettered choices that follow. Choose the answer which you consider correct or most nearly correct. Mark the answer sheet for the letter you have chosen: A, B, C, or D. Check your answers with the correct answers at the end of this chapter.

1. The maximum number of 120-pound weights which can be safely lifted with a chain hoist of 1000 pounds capacity is
   (A) 7
   (B) 8
   (C) 9
   (D) 10

2. The number 0.045 can also be expressed as
   (A) 45 tenths
   (B) 45 hundredths
   (C) 45 thousandths
   (D) 45 ten-thousandths

3. If one gallon of insecticide is needed to spray an area of 1200 square feet, the amount of insecticide needed for a 32-square yard area is most nearly
   (A) 1/2 pint
   (B) 1 pint
   (C) 1 quart
   (D) 1/2 gallon

4. If the recommended amount of napthalene for use in closets is one ounce for each six cubic feet, then the amount needed for a closet 6 feet by 8 feet by 4 feet is

   (A) 1 pound
   (B) 2 pounds
   (C) 3 pounds
   (D) 4 pounds

5. Assume that a foreman has assigned six (6) men to do a certain job in four (4) days. Due to an emergency he can assign only two (2) men. Assuming that these two men work at the same speed, how many days will it take them to complete the job?
   (A) 1 1/3 days
   (B) 12 days
   (C) 16 days
   (D) 18 days

6. Assume that you, the foreman, must assign exterminators to service in one week all apartments in a housing development. Exterminators are scheduled to work five seven-hour days in a week. If one hundred and forty hours are required to service this development, how many exterminators should be assigned?
   (A) one
   (B) two
   (C) three
   (D) four

7. The sum of 1'-9 3/4", 0'-2 7/8", 3'-0", 4'-6 3/8", and 7'-2 3/4" is
   (A) 16'-9 3/4"
   (B) 16'-9 1/4"
   (C) 16'-8 3/4"
   (D) 16'-8 1/4"

8. A crate contains 3 pieces of equipment weighing 73, 84, and 47 pounds respectively. If the crate is lifted by 4 men, each lifting one corner of the crate, the average number of pounds lifted by each of the men is
   (A) 51
   (B) 55
   (C) 61
   (D) 68

9. A slab of concrete is 126 feet long, 36 feet wide and 9 inches thick. Its volume, in cubic yards is most nearly
   (A) 28 yd.³
   (B) 42 yd.³
   (C) 126 yd.³
   (D) 84 yd.³

10. A pipe is laid on an upward slope of 1/4" vertical for each one-foot horizontal. In a horizontal distance of 30 feet, the vertical distance the pipe rises, is
    (A) 6 7/8"
    (B) 7 1/2"
    (C) 7 7/8"
    (D) 8 1/2"

11. A distance of 32 7/8" is most nearly
    (A) 2.54 ft.
    (B) 2.64 ft.
    (C) 2.74 ft.
    (D) 2.84 ft.

12. The sum of numbers 5'-10 7/8", 17'-1/2", 22'-7 1/16" is
    (A) 44'-5 5/16"
    (B) 45'-6 7/16"
    (C) 45'-8 1/16"
    (D) 46'-3 3/16"

13. The difference between 45'-6 1/2" and 27'-8 3/4" is
    (A) 18'-4 1/4"
    (B) 18'-2 3/4"
    (C) 17'-11 1/4"
    (D) 17'-9 3/4"

14. The area of a circle 8 feet in diameter is most nearly
    (A) 45 ft.²
    (B) 50 ft.²
    (C) 55 ft.²
    (D) 60 ft.²

15. Sixty-two fittings, each weighing 268 pounds, are used on a job. If one ton equals 2,000 pounds, the total weight of the sixty-two fittings is most nearly
    (A) 7.5 tons
    (B) 7.9 tons
    (C) 8.3 tons
    (D) 8.7 tons

16. Five and one-quarter percent of $8,752.00 is
    (A) $457.35
    (B) $458.26
    (C) $459.48
    (D) $460.50

17. The jaws of a vise close 3/16" for each turn of the screw. If the vise is open 3 inches, the number of turns needed to close the jaws completely is
    (A) 15
    (B) 16
    (C) 17
    (D) 18

18. A certain job, which took 12 days of 8 hours each to complete, required 6 workers at $4.80 per hour, and a foreman whose salary is equivalent to $45.00 per day. The total labor cost for the job was
    (A) $885.60   (C) $2764.80
    (B) $1000.80   (D) $3304.80

19. A revolution counter applied to the end of a rotating shaft reads 200 when a stop-watch is started and 950 after 90 seconds. The shaft is rotating at a speed of
    (A) 500 RPM   (C) 750 RPM
    (B) 575 RPM   (D) 1150 RPM

20. Five equally spaced holes of one-inch diameter are to be drilled in a straight line in a steel plate. The distance between the centerlines of the end holes is 12 inches. The distance between the centerlines of two adjacent holes is
    (A) 1.00 inch   (C) 2.40 inches
    (B) 2.00 inches   (D) 3.00 inches

TEST IV. SHOP ARITHMETIC

DIRECTIONS: For each question in this test, read carefully the stem and the four lettered choices that follow. Choose the answer which you consider correct or most nearly correct. Mark the answer sheet for the letter you have chosen: A, B, C, or D. Check your answers with the correct answers at the end of this chapter.

1. If a train on a certain route makes two round trips in 5 hours and 20 minutes, the average time for one round trip would be, in minutes
   (A) 100 minutes
   (B) 150 minutes
   (C) 160 minutes
   (D) 200 minutes

2. If 3 pieces of wood, each 10 feet long, are to be cut into 3 foot lengths, the maximum number of 3-foot pieces of wood which can be obtained will be
   (A) 6
   (B) 8
   (C) 9
   (D) 10

3. A box contains 3 pieces of equipment weighing 43, 65 and 84 pounds respectively. If the crate is lifted by 4 men, each lifting one corner of the box, the average number of pounds lifted by each of the men is
   (A) 32
   (B) 40
   (C) 48
   (D) 64

4. The sum of 5 feet-1 1/2 inches, 6 feet-2 3/4 inches, and 6 3/4 inches is
   (A) 11'-10 1/4"
   (B) 11'-11"
   (C) 12'-0"
   (D) 12'-2"

5. The material pass for a truck load of pipe valves refers to "one gross". The maximum number of valves permitted to go out of the yard in this truck is
   (A) 12
   (B) 72
   (C) 144
   (D) 240

6. The fraction 3/8 expressed as a decimal is
   (A) 0.125
   (B) 0.250
   (C) 0.333
   (D) 0.375

7. A pound of a certain type of metal washer contains 360 washers. If 1/4 of the material of each washer is removed by enlarging the center of each washer, the number of washers to the pound should then be most nearly
   (A) 280
   (B) 300
   (C) 380
   (D) 480

8. A worker earns $5.42 per hour, and time and one-half for overtime. Ten per cent of his total salary earned is deducted from his pay check for Social Security and taxes. He also contributes $2.50 per week to a charitable organization. No other deductions are made. If he works 2 hours over his basic 40 hours, his weekly take home pay should be most nearly
   (A) $233.06
   (B) $209.75
   (C) $207.25
   (D) $205.30

9. Assume that the average life of a pair of subway car wheels is 100,000 car-miles, and that 15,000 car miles are lost each time the wheels are turned in the lathe to make them suitable for use. If a certain pair of wheels has been sent to the shop for turning, after 40,000 car-miles and again after 65,000 car-miles of operation, then the number of car-miles of operation remaining in the pair of wheels after the second wheel turning is
   (A) 0
   (B) 5,000
   (C) 20,000
   (D) 35,000

10. A car part costs $130 per 100 units if purchased from a vendor. The car part can be made on a machine which can be purchased for $1000. Assume that this machine has a production life of 20,000 units with no salvage value, and that all shop costs amount to $80 per 100 units turned out in the shop. The money that would be saved during the life of the machine would be
    (A) $ 800
    (B) $8,000
    (C) $9,000
    (D) $18,000

11. The budget for a work unit was $1000.00 in 1971. Their 1972 budget was 5% higher than that of 1971, and their 1973 budget was 10% higher than that of 1972. The work unit's budget for 1973 is
    (A) $1,055
    (B) $1,115
    (C) $1,155
    (D) $1,205

12. Three pieces of machinery were recently purchased. One machine cost $1,739.55, the second machine cost $6,284.00. The total cost for all three machines was $12,721.00. How much did the third machine cost?
    (A) $4,607.55
    (B) $4,697.45
    (C) $4,797.55
    (D) $4,798.45

13. An emergency sanitation aide is paid at the rate of $2.40 per hour. He worked 45 hours in one week and was paid double time for 3 of the 45 hours worked during this week. What was his total gross earnings for the week?
    (A) $112.30
    (B) $115.20
    (C) $126.30
    (D) $155.20

14. A metal sheet is to have a row of ten holes drilled parallel to its length. The holes are to be 3 1/2" in diameter and spaced 5 1/2" on centers. The distance from the edge of each end hole to the adjacent edge of the sheet is to be the same distance as the distance between the edges of the other holes. Based on this information the sheet

should have a minimum length of
    (A) 4'-8 3/4"
    (B) 4'-9"
    (C) 5'-6 1/2"
    (D) 5'-9 3/4"

15. It is required that a 1 3/4-inch-diameter shaft be machined to within a tolerance of plus or minus two-thousandths of an inch. The machined shaft will have to be rejected if it has a diameter of
    (A) 1.746 inches
    (B) 1.748 inches
    (C) 1.750 inches
    (D) 1.752 inches

16. A car part made by a manufacturer "X" has a purchase cost of $7,500, and a life of 5 years; it requires a yearly maintenance cost of $50. Manufacturer "Y" offers a similar part of this type for $4,800, with a life of 3 years and a yearly maintenance cost of $75. By purchasing the part offering a better overall value, the yearly savings per unit purchased would be
    (A) $115
    (B) $125
    (C) $135
    (D) $140

17. A car part costs $150 per 50 units when purchased in a finished condition from a vendor. The car part can be made in the shop at a total cost of $2.20 per unit, when made on a machine which can be purchased for $1000. The minimum number of parts which must be made on this machine before the savings equal the cost of the machine is
    (A) 850
    (B) 1000
    (C) 1250
    (D) 1500

18. A shim pack is to be assembled having an overall thickness between the limits of 0.250" and 0.253". If individual shims are available in thicknesses of 0.005", 0.014" and 0.016", the minimum number of shims required to make up the assembly in any combination is
    (A) 18
    (B) 17
    (C) 16
    (D) 15

19. A subway car wheel wears such that the diameter is reduced from 34 inches to 33 1/2 inches after 150,000 miles of operation. The rate of wear of the car wheel on the diameter is
    (A) 1/150 of an inch per 1,000 miles
    (B) 1/6 of an inch per 100,000 miles
    (C) 5/6 of an inch per 250,000 miles
    (D) 4/5 of an inch per 500,000 miles

20. A car part can be overhauled at the rate of 12 parts per hour. Each part requires new material costing $3 each. If the labor cost is $7 per hour, one part can be overhauled for a total cost (labor plus material) of most nearly
    (A) $3.32
    (B) $3.58
    (C) $3.73
    (D) $4.10

Shop Arithmetic—Test I

Correct Answers

1. B	7. C	13. A	19. B	25. B	31. A
2. D	8. A	14. B	20. A	26. A	32. A
3. B	9. C	15. A	21. C	27. D	33. B
4. A	10. C	16. A	22. D	28. C	34. C
5. B	11. C	17. B	23. C	29. D	
6. D	12. D	18. C	24. B	30. C	

Explanatory Answers

1. B $\dfrac{ft.^2}{gal.}$ $\dfrac{400}{1} = \dfrac{73(15)}{x}$

    $400x = 1095$
    $x = 2.7375$
    Use 2¾ gallon.

2. D

    45 m.p.h →

    a ——————————————— b
    1540 mi. ←
    25 m.p.h.

    Let t = time traveled
    $45t + 25t = 1540$
    $70t = 1540$
    $t = 22$ hours

3. B $\dfrac{yd.^3}{sec.}$ $\dfrac{2}{40} = \dfrac{x}{45(60)}$

    $40x = 5400$
    $x = 135$ yds.^{3}

4 to 7

	Inspector	Machine	Paint	Truck	Total
ft.^{2}	3300	1700	1500	2000	8500
heat cost	Quest. 4	Quest. 5	Quest. 6	Quest 7	1870

4. A $\dfrac{3300}{8500}(1870) = \$726$

5. B $\dfrac{1700}{8500}(1870) = \$374$

6. D $\dfrac{1500}{8500}(1870) = \$330$

7. C $\dfrac{2000}{8500}(1870) = \$440$

8. A 1 yd.^{2} = 9 ft.^{2}
    $36(9) = 324$ ft.^{2}

9. C

    h = 5
    3
    4

    $(Leg)^2 + (Leg)^2 = (Hyp)^2$
    $3^2 + 4^2 = h^2$
    $25 = h^2$
    $h = 5$ feet

10. C Length = $3^5/_{16} + 2^3/_4$
Length = $3^5/_{16} + 2^{12}/_{16}$
Length = $5^{17}/_{16}$
Length = $6^1/_{16}$ inches

11. C

There are 16 sheets of 8×11 in 1 sheet of 32×44. 4000 sheets of $8 \times 11 = \dfrac{4000}{16} = 250$ sheets of 32×44.

$$\frac{500}{\$12} = \frac{250}{x}$$

12. D $$\frac{\$120}{100} = \frac{x}{20}$$

$100x = \$2400$
$x = \$24$

13. A   5 ft.-2¾ in.
+ 8 ft.- ½ in.
+   12½ in.
────────────
13 ft.-15¾ in. $\longrightarrow$ 14 feet, 3¾ inches

14. B $d = rt$
$1 = 30t$
$t = 1/30$ hr.
$t = 2$ minutes

15. A $$\frac{6 \text{ sacks}}{1 \text{ yd.}^3} = \frac{x}{(4 \text{ in.})(6 \text{ ft.})(27 \text{ ft.})}$$

$$\frac{6}{1} = \frac{x}{\left(\frac{4}{36}\right)\left(\frac{6}{3}\right)\left(\frac{27}{3}\right)}$$

$$x = 6\left(\frac{4}{36}\right)\left(\frac{6}{3}\right)\left(\frac{27}{3}\right)$$

$x = 12$ sacks

16. A Bolt $= \dfrac{1}{6}x$; Nut $= \dfrac{1}{30}x$

$$\frac{x}{6} + \frac{x}{30} = 60$$

$$\frac{5x + x}{30} = 60$$

$6x = 1800$
$x = 300$ nuts

17.   B $$\frac{.25 \text{ lb.}}{1 \text{ in.}^3} = \frac{x}{(8 \text{ ft.-6 in.}) (4 \text{ in.}) (\frac{1}{2} \text{ in.})}$$

$$\frac{\frac{1}{4}}{1} = \frac{x}{(102) (4) (\frac{1}{2})}$$

$$x \quad = \frac{1}{4} (102) (4) (\frac{1}{2})$$
$$x \quad = 51 \text{ lbs.}$$

18.   C $(8) (5) (9.75) = \$390$

19.   B   Step One:
$$C = 2\pi r$$
$$\text{diameter} = 1 \text{ ft.-9 in.}$$
$$= 21 \text{ in.}$$
$$\text{radius} = 2\frac{1}{2} \text{ in.}$$
$$C = 2\pi (2\frac{1}{2})$$

Step Two:
$$C = 2\pi (2\frac{1}{2})$$
$$C = 2(2\frac{2}{7}) (2\frac{1}{2})$$
$$C = 66''$$

20.   A   diameter $= 1$ inch; radius $= \frac{1}{2}$ inch
$$A = \pi r^2$$
$$A = (2\frac{2}{7}) (\frac{1}{2})^2$$
$$A = 2\frac{2}{28} \approx \frac{3}{4}$$
$$A = \frac{3}{4} \text{ square inch}$$

21.   C
$$SA = 6e^2$$
$$SA = 6(5)^2$$
$$SA = 150 \text{ square inches}$$

22.   D $27 \text{ ft.}^3 = 1 \text{ yd.}^3$

23.   C   The difference between a 6 gallon and 4 gallon can is 2 gallons.

24.   B $$\frac{3 \text{ rooms}}{(6 \text{ men}) (4 \text{ hrs.})} = \frac{2 \text{ rooms}}{(4 \text{ men}) (x \text{ hrs.})}$$

$$12x = 48$$
$$x = 4 \text{ hours}$$

25.   B $\frac{1}{2}$ ton $= \frac{1}{2} (2000 \text{ lbs.}) = 1000 \text{ lbs.}$

$$\frac{1000}{125} = 8 \text{ weights}$$

26.   A   Diameter is 14; radius is 7.
              Length = 30 [2πr]
              Length = 30 [2($^{22}/_7$)(7)]
              Length = 30(44)
              Length = 1320 in.
              Length = 110 ft.

27.   D   No. 9 weighs 40 lbs./1000 ft.; No. 12 weighs 40 lbs./2000 ft.

              $$\frac{40}{2000} = \frac{x}{500}$$

              x = 10 lbs.

28.   C $$\frac{.32 \text{ lbs.}}{1 \text{ in.}^3} = \frac{x}{(6 \text{ ft.}) (\frac{1}{4} \text{ in.}) (1 \text{ in.})}$$

              $$\frac{.32}{1} = \frac{x}{[6 (12)](\frac{1}{4}) (1)}$$

              x = .32(72) (¼)
              x = 5.76 pounds

29.   D   Step One:
                Saves   4.1
                      − 1.7
                      ‾‾‾‾‾
                        2.4 miles/day

              Step Two:
                Over 129 days = 2.4 (129) = 309.6 miles saved

              Step Three:
                (miles saved) (time spent on each mile)
                309.6 miles $\left(\dfrac{73 \text{ min.}}{4.1 \text{ miles}}\right)$ = 5512 min.
                                              = 91.8 hrs.

30.   C   (.85) (340) (10) + .15 (340) = 2890 + 51 = 2941 lbs.

31.   A   1.51 − .015 < x < 1.51 + .015
                    1.495 < x < 1.565
                    x is greater than 1.490″

32.   A $$\frac{1250}{5} = \frac{x}{20}$$

              $$x = \frac{20(1250)}{5}$$

              x = 5000 bricks

33.   B   (25 watt) (100 hrs.) : (100 watt) (25 hrs.)
                          2500 : 2500

34.   C $^7/_{32}$ ≈ .219 inches

Shop Arithmetic—Test II

Correct Answers

1. B	4. D	7. C	10. D	13. B	16. D	19. B
2. C	5. B	8. C	11. C	14. B	17. B	20. B
3. C	6. C	9. D	12. B	15. B	18. B	21. C

Explanatory Answers

1. B

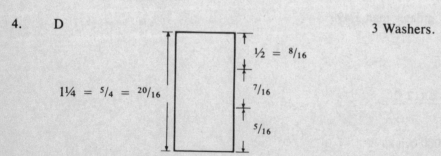

Maximum is 15 gaskets.

2. C

$5^2 - 3^2 = 16$ in.^{2}

3. C

$$\frac{7}{\frac{7}{8}} = \frac{x}{\frac{8}{8}}$$

$$7 = \frac{7}{8}x$$

$$x = \frac{56}{7}$$

$$x = 8 \text{ threads}$$

4. D

3 Washers.

$$\frac{1}{2} = \frac{8}{16}$$

$$1\frac{1}{4} = \frac{5}{4} = \frac{20}{16}$$

$$\frac{7}{16}$$

$$\frac{5}{16}$$

5. B   The movement on one pipe is compensated for by the movement on the other pipe.

6. C   Volumes are equal. The least height has the greatest base area.

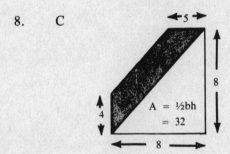

No. 1   No. 2   No. 3   No. 4
A = bh = 22   A = bh = 27   A = bh = 28   A = s^2 = 25

7. C   V = ($^3/_2$) (2) (8) + (¾) (2) (16)
        V = 24 + 24
        V = 48 in.^{3}

8. C   A = ½bh
        = 32

        A = 32 in.^{2}

9. D   V = (4) (8) (2) in.^{3}
        V = 64 in.^{3}
        V = $\dfrac{64}{(12)(12)(12)}$
        V = $\dfrac{1}{27}$ ft.^{3}

10. D

        RPM varies inversely to the number of teeth.
            120 (24) = (RPM) (12)
                RPM = 240

        24 teeth   36 teeth   12 teeth

11. C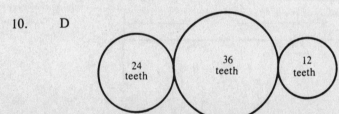

        x = ½ + ½ + 7¼ + 7¼ + 7¼ + ½ + ½
        x = 2 + 21¾
        x = 23¾"

12. B   Average = $\dfrac{\text{Sum of items}}{\text{Number of items}}$

        Average = $\dfrac{4 + 4 + 3 + 4 + 10}{5}$

        Average = $\dfrac{25}{5}$

        Average = 5 feet

13.   B   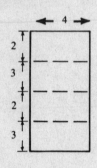   Diameter is $^7/_2$; radius is $^7/_4$

2 + 2 + Circumference
4 + 2($^{22}/_7$) ($^7/_4$)
4 + 11 = 15
Length = 15″

14.   B   13 threads.

15.   B   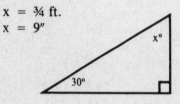   A = 8 + 12 + 8 + 12 = 40
A = 40 + 2½% of 40
A = 1.025(40)
A = 41 ft.²

16.   D $\dfrac{1 \text{ inch}}{1 \text{ ft.}} = \dfrac{¾ \text{ inch}}{x}$

x = ¾ ft.
x = 9″

17.   B   x + 30 + 90 = 180°
x = 60°

18.   B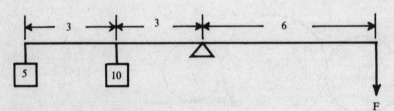

clockwise moments = counter-clockwise moments

(f) (d) = (f) (d)
6F = 3 (10) + 6 (5)
6F = 60
F = 10 lbs.

19.   B   1 inch : 10 feet
(1 inch)² : (10 feet)² → 1 in.² = 100 ft.²
A square 5 in. × 5 in. has 25 in.²

$\dfrac{1 \text{ in.}^2}{100 \text{ ft.}^2} = \dfrac{25 \text{ in.}^2}{x}$

x = 2500 ft.²

20.   B

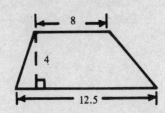

$$A = h \left(\frac{b_1 + b_2}{2} \right)$$

$$A = 4 \left(\frac{8 + 12.5}{2} \right)$$

$$A = 4 \left(\frac{20.5}{2} \right)$$

$$A = 41$$

21.   C

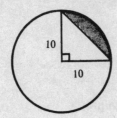

Shaded area = ¼ circle area minus area of right triangle.

$$= \frac{1}{4}(\pi r^2) - \frac{1}{2}bh$$

$$= \frac{1}{4}(^{22}/_7)(10)^2 - \frac{1}{2}(10)(10)$$

$$= 78.57 - 50 \approx 28.5$$

Shaded area ≈ 28.5 ft.²

Shop Arithmetic—Test III

Correct Answers

1. B	5. B	9. C	13. D	17. B
2. C	6. D	10. B	14. B	18. D
3. C	7. A	11. C	15. C	19. A
4. B	8. A	12. B	16. C	20. D

Explanatory Answers

1.   B $\frac{1000}{125}$ = 8 weights

2.   C $.045 = \frac{45}{1000}$ = 45 thousandths

3.   C $\frac{1 \text{ gallon}}{1200 \text{ ft.}^2} = \frac{x}{32 \text{ yd.}^2} = \frac{x}{32(9) \text{ ft}^2}$

$$\frac{1}{1200} = \frac{x}{32(9)}$$

$$1200x = 288$$
$$x = .24 \approx \frac{1}{4} \text{ gallon}$$
$$x \approx 1 \text{ quart}$$

4. B $\dfrac{1 \text{ oz.}}{6 \text{ ft.}^3} = \dfrac{x}{6\,(8)\,(4)}$

$$6x = 6\,(8)\,(4)$$
$$x = 32 \text{ oz.}$$
$$x = 2 \text{ pounds}$$

5. B $\dfrac{1 \text{ job}}{(6 \text{ men})\,(4 \text{ days})} = \dfrac{1 \text{ job}}{(2 \text{ men})\,(x \text{ days})}$

$$\frac{1}{24} = \frac{1}{2x}$$

$$x = 12 \text{ days}$$

6. D Exterminator has 5 (7) = 35 hrs., needs 140 hrs.
$$\frac{140}{35} = 4$$

7. A
$$
\begin{array}{rl}
1 \text{ ft.} - & 9.75'' \\
 & 2.875'' \\
3 \text{ ft.} & \\
4 \text{ ft.} - & 6.375'' \\
+\ 7 \text{ ft.} - & 2.75'' \\
\hline
15 \text{ ft.} - & 21.75'' \rightarrow 16'\text{-}9¾''
\end{array}
$$

8. A $\dfrac{73 + 84 + 47}{4} = \dfrac{204}{4} = 51 \text{ pounds}$

9. C $(126 \text{ ft.})\,(36 \text{ ft.})\,(9 \text{ in.}) = \left(\dfrac{126}{3}\right)\left(\dfrac{36}{3}\right)\left(\dfrac{9}{36}\right) \text{ yd.}^3 = 126 \text{ yd.}^3$

10. B

$\dfrac{12}{360} = \dfrac{¼}{x}$

$$12x = 90$$
$$x = 7½''$$

11. C $32⅞ \text{ in.} = \dfrac{32.875}{12} \approx 2.74 \text{ ft.}$

12. B
$$
\begin{array}{rl}
5 \text{ ft.} - & 10.875 \text{ in.} \\
17 \text{ ft.} - & .5 \ \text{ in.} \\
+\ 22 \text{ ft.} - & 7.0625 \text{ in.} \\
\hline
44 \text{ ft.} - & 18.4375 \text{ in.} \rightarrow 45 \text{ ft.-}6^7/_{16} \text{ in.}
\end{array}
$$

13.   D

$$
\begin{array}{l}
\quad 45 \text{ ft. - 6.5 in.} \\
\underline{- 27 \text{ ft. - 8.75 in.}}
\end{array} \rightarrow
\begin{array}{l}
\quad 44 \text{ ft. - 18.5 in.} \\
\underline{- 27 \text{ ft. - 8.75 in.}} \\
\quad 17 \text{ ft. - 9.75 in.} = 17 \text{ ft. - 9¾ in.}
\end{array}
$$

14.   B

$$A = \pi r^2$$
$$A = \frac{22}{7}(4^2) \approx 50$$
$$A \approx 50 \text{ ft.}^2$$

15.   C

$$\frac{62(268)}{2000} = \frac{16616}{2000} = 8.3 \text{ tons}$$

16.   C

$$(5¼\%)(8752) = (.0525)(8752) = \$459.48$$

17.   B

$$\frac{3}{\frac{3}{16}} = \frac{48}{3} = 16 \text{ turns}$$

18.   D

$$
\begin{array}{ccccc}
& \text{workers} & \text{hours} & \text{rate} & \begin{array}{c}\text{foreman}\\\text{rate}\end{array} & \text{days}
\end{array}
$$

Cost =   (6)   ((12) (8)) (4.80) + (45) (12)
Cost = 2764.8 + 540
Cost = \$3304.80

19.   A

$$\frac{950 - 200 \text{ rev.}}{90 \text{ sec.}}$$

$$= \frac{750 \text{ rev.}}{90 \text{ sec.}} \cdot \frac{60 \text{ sec.}}{1 \text{ min.}}$$

$$= 500 \frac{\text{rev.}}{\text{min.}}$$

$$= 500 \text{ RPM}$$

20.   D   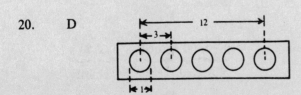   3.00 inches

Shop Arithmetic—Test IV

Correct Answers

1. C	5. C	9. B	13. B	17. C
2. C	6. D	10. C	14. B	18. C
3. C	7. D	11. C	15. A	19. C
4. B	8. C	12. B	16. B	20. B

Explanatory Answers

1. C   5 hr. - 20 min. = 320 min. $\dfrac{320}{2}$ = 160 minutes

2. C

Three 3 ft. pieces per board.
(3 pieces per board) (3 boards) = 9 pieces

3. C $\dfrac{43 + 65 + 84}{4} = \dfrac{192}{4}$ = 48 lbs.

4. B
$$
\begin{array}{r}
5 \text{ ft. } - 1.5'' \\
6 \text{ ft. } - 2.75'' \\
+\qquad 6.75'' \\
\hline
11 \text{ ft. } - 11''
\end{array}
$$

5. C   1 gross = 144 valves

6. D $\dfrac{3}{8}$ = .375

7. D   If remove ¼ material then $\dfrac{¾ \text{ lb.}}{360} = \dfrac{1 \text{ lb.}}{x}$

$x = 360 \left(\dfrac{4}{3}\right)$

$x = 480$ washers

8. C $\underbrace{(5.42) \text{ (hrs.)} + (1.5) (5.42) \text{ (overtime hrs.)}}_{\text{total earned}} - \underbrace{(10\% \text{ total earned})}_{\text{S.S. and taxes}} \quad \underbrace{2.50}_{\text{charity}}$

works 40 hours + 2 hours overtime

$\underbrace{(5.42) (40) + (1.5) (5.42) (2)}_{233.06} - (.10) (233.06) - 2.50 \approx 207.25$

9. B 100,000
  − 40,000 used
  − 15,000 turning
  − 25,000 additional used
  − 15,000 turning
     ─────────────
     5,000

10. C Vendor: $\dfrac{130}{100 \text{ units}}$ (20,000 units) = 26,000

  Machine: 1000 + $\dfrac{80}{100 \text{ units}}$ (20,000 units) = 17,000

  $26,000
  −$17,000
  ─────────
  $ 9,000

11. C
  1971 1972
  1.05 (1000) = 1050

  1972 1973
  1.1 (1050) = 1155 = $1,155

12. B
  1,739.55   12,721
  + 6,284   − 8,023.55
  ──────────   ──────────
  $ 8,023.55 $ 4,697.45

13. B (45 − 3) (2.40) + 3 [2(2.40)] = $115.20

14. B

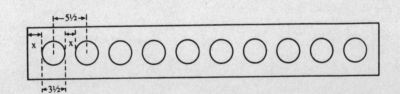

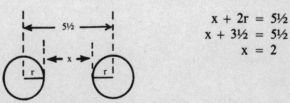

$$x + 2r = 5\tfrac{1}{2}$$
$$x + 3\tfrac{1}{2} = 5\tfrac{1}{2}$$
$$x = 2$$

Length = center to center first hole to last
  + half hole at each end
  + edge distance at each end

$$L = 9 \, (5\tfrac{1}{2}) + 2 \left(\dfrac{3\tfrac{1}{2}}{2}\right) + 2(2)$$

$$L = 57 \text{ in.} = \frac{57}{12} \text{ ft.}$$

$$L = 4 \text{ ft.} - 9 \text{ in.}$$

$$L = 4' - 9''$$

15.   A $1.75 - .002 \leq x \leq 1.75 + .002$

$1.748 \leq x \leq 1.752$

1.764" is not between these values.

16.   B   x: $\dfrac{7500}{5 \text{ yrs.}} + 50 = 1550/\text{yr.}$

y: $\dfrac{4800}{3 \text{ yrs.}} + 75 = 1675/\text{yr.}$

x saves \$125/yr.

17.   C   Vendor: $\dfrac{150}{50 \text{ units}}$; Machine: $1000 + \dfrac{2.20}{1 \text{ unit}}$

Let x = the number of units

$$\frac{150}{50} (x) = 1000 + \frac{2.20}{1} (x)$$

$$\begin{aligned} 3x &= 1000 + 2.2x \\ .8x &= 1000 \\ x &= 1250 \text{ parts} \end{aligned}$$

18. C

$$.250 - .253 \left\{ \begin{array}{l} \text{use} \\ 14(.016) = .224 \\ 2(.014) = \dfrac{.028}{.252} \end{array} \right. \quad \text{16 pieces.}$$

19. C $\dfrac{\frac{1}{2}\text{ in.}}{150,000\text{ mi.}} = \dfrac{1}{300,000}$

A. $\dfrac{\frac{1}{150}}{1000} = \dfrac{1}{150,000}$

B. $\dfrac{\frac{1}{6}}{100,000} = \dfrac{1}{600,000}$

C. $\boxed{\dfrac{\frac{5}{6}}{250,000} = \dfrac{1}{300,000}}$

D. $\dfrac{\frac{4}{5}}{500,000} = \dfrac{1}{625,000}$

20. B $\dfrac{\$}{\text{part}}$

$\dfrac{\$7}{\text{hr.}} \cdot \dfrac{1\text{ hr.}}{12\text{ parts}} = \dfrac{\$7}{12\text{ parts}} \approx .58 + 3.00 \text{ mat} = \$3.58$